POCKET ON NEXT PAGE

Main–Group Elements

10	11 I B	12 II B	13 III A	14 IV A	15 V A	16 VI A	17 VII A	18 VIII A
								2 4.002602 He helium
			5 10.811 B boron	6 12.011 C carbon	7 14.00674 N nitrogen	8 15.9994 O oxygen	9 18.9984032 F fluorine	10 20.1797 Ne neon
			13 26.981539 Al aluminum	14 28.0855 Si silicon	15 30.973762 P phosphorus	16 32.066 S sulfur	17 35.4527 Cl chlorine	18 39.948 Ar argon
28 58.69 Ni nickel	29 63.546 Cu copper	30 65.39 Zn zinc	31 69.723 Ga gallium	32 72.61 Ge germanium	33 74.92159 As arsenic	34 78.96 Se selenium	35 79.904 Br bromine	36 83.80 Kr krypton
46 106.42 Pd palladium	47 107.8682 Ag silver	48 112.411 Cd cadmium	49 114.82 In indium	50 118.710 Sn tin	51 121.75 Sb antimony	52 127.60 Te tellurium	53 126.90447 I iodine	54 131.29 Xe xenon
78 195.08 Pt platinum	79 196.96654 Au gold	80 200.59 Hg mercury	81 204.3833 Tl thallium	82 207.2 Pb lead	83 208.98037 Bi bismuth	84 (209) Po polonium	85 (210) At astatine	86 (222) Rn radon
110 (269) Uun ununnilium	111 (272) Uuu unununium	112 (277) Uub ununbiium						

Inner–Transition Metals

63 151.965 Eu europium	64 157.25 Gd gadolinium	65 158.92534 Tb terbium	66 162.50 Dy dysprosium	67 164.93032 Ho holmium	68 167.26 Er erbium	69 168.93421 Tm thulium	70 173.04 Yb ytterbium	71 174.967 Lu lutetium
95 (243) Am americium	96 (247) Cm curium	97 (247) Bk berkelium	98 (251) Cf californium	99 (252) Es einsteinium	100 (257) Fm fermium	101 (258) Md mendelevium	102 (259) No nobelium	103 (262) Lr lawrencium

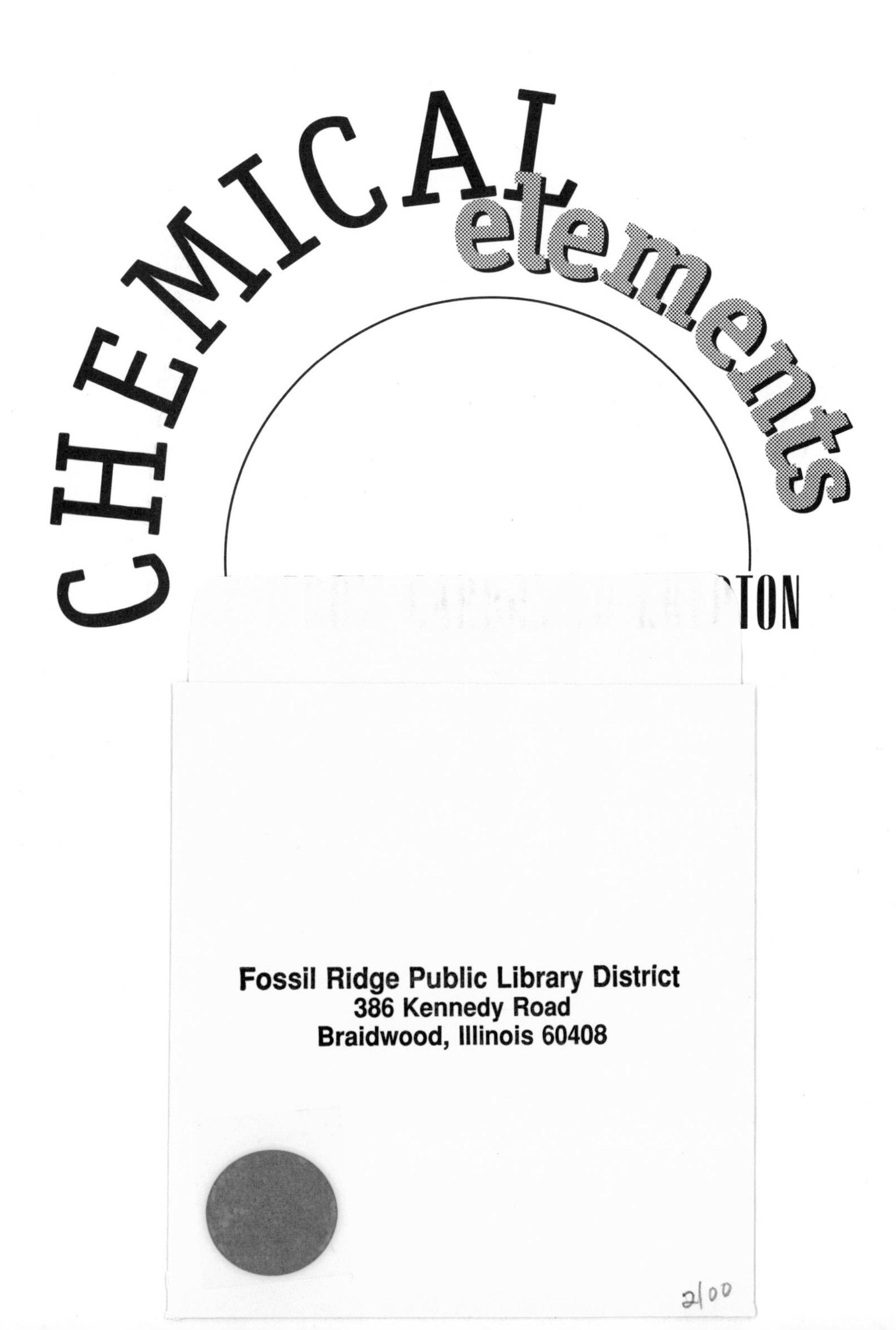
CHEMICAL elements
Fossil Ridge Public Library District
386 Kennedy Road
Braidwood, Illinois 60408
2/00

David E. Newton

Lawrence W. Baker, editor

U·X·L®
AN IMPRINT OF GALE
DETROIT • LONDON

Dedicated to three very special women in my life,
Agnes Greenhaw, Donna Krouse-Slesnick, and Ruth Ranen
"Thanks" is not enough for all you have done for me!

Chemical Elements: From Carbon to Krypton

David E. Newton

Staff

Lawrence W. Baker, *U•X•L Senior Editor*
Carol DeKane Nagel, *U•X•L Managing Editor*
Thomas L. Romig, *U•X•L Publisher*

Shalice Shah-Caldwell, *Permissions Associate (Pictures)*
Jessica L. Ulrich, *Permissions Associate (Pictures)*
Margaret Chamberlain, *Permissions Specialist (Pictures)*

Deborah Milliken, *Production Assistant*
Evi Seoud, *Assistant Production Manager*
Mary Beth Trimper, *Production Director*

Eric Johnson, *Art Director*
Cynthia Baldwin, *Product Design Manager*

Pamela Reed, *Photography Coordinator*
Mike Logusz, *Imaging Specialist*
Randy A. Bassett, *Image Database Supervisor*
Barbara J. Yarrow, *Graphic Services Manager*

Marco Di Vita, The Graphix Group, *Typesetting*

Library of Congress Cataloging-in-Publication Data

Newton, David E.
Chemical elements : from carbon to krypton / David E. Newton, Lawrence W. Baker, editor.
p. cm.
Includes bibliographical references and index.
Contents: v. 1. A-F -- v. 2. G-O -- v. 3. P-Z.
ISBN 0-7876-2844-1 (set). -- ISBN 0-7876-2845-X (v. 1). -- ISBN 0-7876-2846-8 (v. 2). -- ISBN 0-7876-2847-6 (v. 3)
1. Chemical elements. I. Baker, Lawrence W.
QD466.N464 1998
546–dc21 98-31207
CIP

Printed in the United States of America

10 9 8 7 6 5 4 3

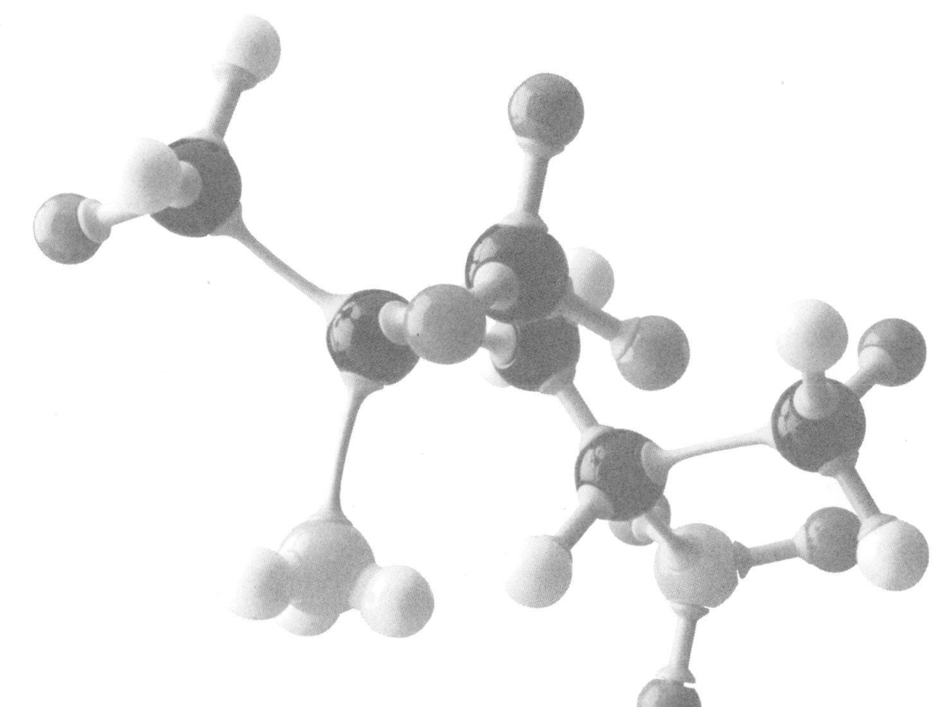

CONTENTS

Contents by Atomic Number xiii
Contents by Family Group xvii
Reader's Guide xxiii
Timeline xxvii
Words to Know xxxvii

Volume 1

Actinium (Ac) 1
Aluminum (Al) 5
Americium (Am) 15
Antimony (Sb) 19
Argon (Ar) 25
Arsenic (As) 31
Astatine (At) 37
Barium (Ba) 41
Berkelium (Bk) 49
Beryllium (Be) 53
Bismuth (Bi) 59
Bohrium (Bh)
See Transfermium elements, volume 3, p. 625

Boron (B) 65
Bromine (Br) 73
Cadmium (Cd) 81
Calcium (Ca) 87
Californium (Cf) 97
Carbon (C) 101
Cerium (Ce) 113
Cesium (Cs) 119
Chlorine (Cl) 125
Chromium (Cr) 135
Cobalt (Co) 141
Copper (Cu) 149
Curium (Cm) 157
Dubnium (Db)
See Transfermium elements, volume 3, p. 625
Dysprosium (Dy) 161
Einsteinium (Es) 165
Erbium (Er) 169
Europium (Eu) 175
Fermium (Fm) 179
Fluorine (F) 183
Francium (Fr) 191

Volume 2

Gadolinium (Gd) 195
Gallium (Ga) 201
Germanium (Ge) 209
Gold (Au) 217
Hafnium (Hf) 225
Hassium (Hs)
See Transfermium elements, volume 3, p. 625
Helium (He) 231
Holmium (Ho) 241
Hydrogen (H) 245
Indium (In) 255
Iodine (I) 261
Iridium (Ir) 269

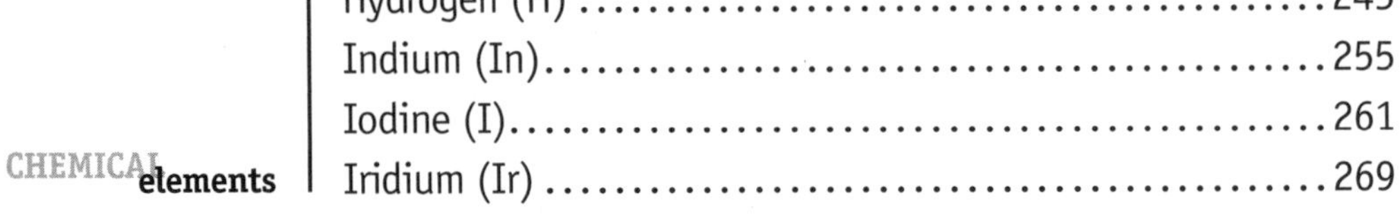

Iron (Fe) 277
Krypton (Kr) 287
Lanthanum (La)...293
Lawrencium (Lr)
See Transfermium elements, volume 3, p. 625
Lead (Pb)...299
Lithium (Li)...307
Lutetium (Lu)...313
Magnesium (Mg) 317
Manganese (Mn) 327
Meitnerium (Mt)
See Transfermium elements, volume 3, p. 625
Mendelevium (Md)
See Transfermium elements, volume 3, p. 625
Mercury (Hg)...333
Molybdenum (Mo)...343
Neodymium (Nd)...349
Neon (Ne) 355
Neptunium (Np)...361
Nickel (Ni)...367
Niobium (Nb) 375
Nitrogen (N) 381
Nobelium (No)
See Transfermium elements, volume 3, p. 625
Osmium (Os)...391
Oxygen (O) 395

Volume 3

Palladium (Pd)...409
Phosphorus (P) 415
Platinum (Pt) 425
Plutonium (Pu) 431
Polonium (Po) 439
Potassium (K)...443
Praseodymium (Pr)...453
Promethium (Pm) 459
Protactinium (Pa) 465

Contents

Radium (Ra) 471
Radon (Rn) 477
Rhenium (Re) 485
Rhodium (Rh) 491
Rubidium (Rb) 495
Ruthenium (Ru) 501
Rutherfordium (Rf)
See Transfermium elements, volume 3, p. 625
Samarium (Sm) 505
Scandium (Sc) 511
Seaborgium (Sg)
See Transfermium elements, volume 3, p. 625
Selenium (Se) 517
Silicon (Si) 525
Silver (Ag) 533
Sodium (Na) 541
Strontium (Sr) 553
Sulfur (S) 559
Tantalum (Ta) 569
Technetium (Tc) 575
Tellurium (Te) 579
Terbium (Tb) 585
Thallium (Tl) 591
Thorium (Th) 597
Thulium (Tm) 605
Tin (Sn) 609
Titanium (Ti) 617
Transfermium elements 625
Tungsten (W) 633
Ununbiium (Uub)
See Transfermium elements, volume 3, p. 625
Ununnilium (Uun)
See Transfermium elements, volume 3, p. 625
Unununium (Uuu)
See Transfermium elements, volume 3, p. 625
Uranium (U) 639
Vanadium (V) 647

Xenon (Xe) 653
Ytterbium (Yb) 659
Yttrium (Y) 663
Zinc (Zn) 671
Zirconium (Zr) 681

Bibliography xlix
Picture Credits lv
Index lxi

CONTENTS

Elements by Atomic Number

Bold-Italic type indicates volume numbers

ATOMIC NUMBER | ELEMENT (SYMBOL) — VOLUME & PAGE NUMBER

1 Hydrogen (H) ***2:*** 245
2 Helium (He) ***2:*** 231
3 Lithium (Li) ***2:*** 307
4 Beryllium (Be) ***1:*** 53
5 Boron (B) ***1:*** 65
6 Carbon (C) ***1:*** 101
7 Nitrogen (N) ***2:*** 381
8 Oxygen (O) ***2:*** 395
9 Fluorine (F) ***1:*** 183
10 Neon (Ne) ***2:*** 355
11 Sodium (Na) ***3:*** 541
12 Magnesium (Mg) ***2:*** 317
13 Aluminum (Al) ***1:*** 5
14 Silicon (Si) ***3:*** 525
15 Phosphorus (P) ***3:*** 415
16 Sulfur (S) ***3:*** 559
17 Chlorine (Cl) ***1:*** 125
18 Argon (Ar) ***1:*** 25

Contents: Elements by Atomic Number

19 Potassium (K) ***3:*** 443
20 Calcium (Ca) ***1:*** 87
21 Scandium (Sc) ***3:*** 511
22 Titanium (Ti) ***3:*** 617
23 Vanadium (V) ***3:*** 647
24 Chromium (Cr) ***1:*** 135
25 Manganese (Mn) ***2:*** 327
26 Iron (Fe) ***2:*** 277
27 Cobalt (Co) ***1:*** 141
28 Nickel (Ni) ***2:*** 367
29 Copper (Cu) ***1:*** 149
30 Zinc (Zn) ***3:*** 671
31 Gallium (Ga) ***2:*** 201
32 Germanium (Ge) ***2:*** 209
33 Arsenic (As) ***1:*** 31
34 Selenium (Se) ***3:*** 517
35 Bromine (Br) ***1:*** 73
36 Krypton (Kr) ***2:*** 287
37 Rubidium (Rb) ***3:*** 495
38 Strontium (Sr) ***3:*** 553
39 Yttrium (Y) ***3:*** 663
40 Zirconium (Zr) ***3:*** 681
41 Niobium (Nb) ***2:*** 375
42 Molybdenum (Mo) ***2:*** 343
43 Technetium (Tc) ***3:*** 575
44 Ruthenium (Ru) ***3:*** 501
45 Rhodium (Rh) ***3:*** 491
46 Palladium (Pd) ***2:*** 409
47 Silver (Ag) ***3:*** 533
48 Cadmium (Cd) ***1:*** 81
49 Indium (In) ***2:*** 255
50 Tin (Sn) ***3:*** 609
51 Antimony (Sb) ***1:*** 19
52 Tellurium (Te) ***3:*** 579
53 Iodine (I) ***2:*** 261
54 Xenon (Xe) ***3:*** 653
55 Cesium (Cs) ***1:*** 119

56 Barium (Ba) *1:* 41
57 Lanthanum (La) *2:* 293
58 Cerium (Ce) *1:* 113
59 Praseodymium (Pr) *3:* 453
60 Neodymium (Nd) *2:* 349
61 Promethium (Pm).................................... *3:* 459
62 Samarium (Sm).................................... *3:* 505
63 Europium (Eu).................................... *1:* 175
64 Gadolinium (Gd).................................... *2:* 195
65 Terbium (Tb) *3:* 585
66 Dysprosium (Dy).................................... *1:* 161
67 Holmium (Ho).................................... *2:* 241
68 Erbium (Er).................................... *1:* 169
69 Thulium (Tm).................................... *3:* 605
70 Ytterbium (Yb) *3:* 659
71 Lutetium (Lu) *2:* 313
72 Hafnium (Hf).................................... *2:* 225
73 Tantalum (Ta) *3:* 569
74 Tungsten (W).................................... *3:* 633
75 Rhenium (Re) *3:* 485
76 Osmium (Os) *2:* 391
77 Iridium (Ir) *2:* 269
78 Platinum (Pt) *3:* 425
79 Gold (Au) *2:* 217
80 Mercury (Hg) *2:* 333
81 Thallium (Tl) *3:* 591
82 Lead (Pb) *2:* 299
83 Bismuth (Bi) *1:* 59
84 Polonium (Po).................................... *3:* 439
85 Astatine (At) *1:* 37
86 Radon (Rn).................................... *3:* 477
87 Francium (Fr).................................... *1:* 191
88 Radium (Ra).................................... *3:* 471
89 Actinium (Ac).................................... *1:* 1
90 Thorium (Th) *3:* 597
91 Protactinium (Pa) *3:* 465
92 Uranium (U).................................... *3:* 639

93	Neptunium (Np)	***2:*** 361
94	Plutonium (Pu)	***3:*** 431
95	Americium (Am)	***1:*** 15
96	Curium (Cm)	***1:*** 157
97	Berkelium (Bk)	***1:*** 49
98	Californium (Cf)	***1:*** 97
99	Einsteinium (Es)	***1:*** 165
100	Fermium (Fm)	***1:*** 179
101	Mendelevium (Md)	***3:*** 625
102	Nobelium (No)	***3:*** 625
103	Lawrencium (Lr)	***3:*** 625
104	Rutherfordium (Rf)	***3:*** 625
105	Dubnium (Db)	***3:*** 625
106	Seaborgium (Sg)	***3:*** 625
107	Bohrium (Bh)	***3:*** 625
108	Hassium (Hs)	***3:*** 625
109	Meitnerium (Mt)	***3:*** 625
110	Ununnilium (Uun)	***3:*** 625
111	Unununium (Uuu)	***3:*** 625
112	Ununbiium (Uub)	***3:*** 625

CONTENTS

Elements by Family Group

Bold-Italic type indicates volume numbers

Group 1 (IA)

Cesium (Cs) ***1:*** 119
Francium (Fr) ***1:*** 191
Hydrogen (H) ***2:*** 245
Lithium (Li) ***2:*** 307
Potassium (K) ***3:*** 443
Rubidium (Rb) ***3:*** 495
Sodium (Na) ***3:*** 541

Group 2 (IIA)

Barium (Ba) ***1:*** 41
Beryllium (Be) ***1:*** 53
Calcium (Ca) ***1:*** 87
Magnesium (Mg) ***2:*** 317
Radium (Ra) ***3:*** 471
Strontium (Sr) ***3:*** 553

Group 3 (IIIB)

Actinium (Ac) ***1:*** 1
Lanthanum (La) ***2:*** 293

Scandium (Sc) *3:* 511
Yttrium (Y) *3:* 663

Group 4 (IVB)

Hafnium (Hf) *2:* 225
Titanium (Ti).. *3:* 617
Rutherfordium (Rf) *3:* 625
Zirconium (Zr) *3:* 681

Group 5 (VB)

Dubnium (Db).. *3:* 625
Niobium (Nb) *2:* 375
Tantalum (Ta) *3:* 569
Vanadium (V) *3:* 647

Group 6 (VIB)

Chromium (Cr) *1:* 135
Molybdenum (Mo).. *2:* 343
Seaborgium (Sg) *3:* 625
Tungsten (W) *3:* 633

Group 7 (VIIB)

Bohrium (Bh) *3:* 625
Manganese (Mn) *2:* 327
Rhenium (Re) *3:* 485
Technetium (Tc).. *3:* 575

Group 8 (VIIIB)

Hassium (Hs) *3:* 625
Iron (Fe).. *2:* 277
Osmium (Os).. *2:* 391
Ruthenium (Ru).. *3:* 501

Group 9 (VIIIB)

Cobalt (Co) *1:* 141
Iridium (Ir) *2:* 269
Meitnerium (Mt) *3:* 625
Rhodium (Rh).. *3:* 491

Group 10 (VIIIB)

Nickel (Ni) *2:* 367
Palladium *3:* 409
Platinum (Pt) *3:* 425
Ununnilium (Uun) *3:* 625

Group 11 (IB)

Copper (Cu) *1:* 149
Gold (Au) *2:* 217
Silver (Ag) *3:* 533
Unununium (Uuu) *3:* 625

Group 12 (IIB)

Cadmium (Cd) *1:* 81
Mercury (Hg) *2:* 333
Ununbiium (Uub) *3:* 625
Zinc (Zn) *3:* 671

Group 13 (IIIA)

Aluminum (Al) *1:* 5
Boron (B) *1:* 65
Gallium (Ga) *2:* 201
Indium (In) *2:* 255
Thallium (Tl) *3:* 591

Group 14 (IVA)

Carbon (C) *1:* 101
Germanium (Ge) *2:* 209
Lead (Pb) *2:* 299
Silicon (Si) *3:* 525
Tin (Sn) *3:* 609

Group 15 (VA)

Antimony (Sb) *1:* 19
Arsenic (As) *1:* 31
Bismuth (Bi) *1:* 59
Nitrogen (N) *2:* 381
Phosphorus (P) *3:* 415

Group 16 (VIA)

Oxygen (O) ***2:*** 395
Polonium (Po) ***3:*** 439
Selenium (Se) ***3:*** 517
Sulfur (S) ***3:*** 559
Tellurium (Te) ***3:*** 579

Group 17 (VIIA)

Astatine (At) ***1:*** 37
Bromine (Br) ***1:*** 73
Chlorine (Cl) ***1:*** 125
Fluorine (F) ***1:*** 183
Iodine (I) ***2:*** 261

Group 18 (VIIIA)

Argon (Ar) ***1:*** 25
Helium (He) ***2:*** 231
Krypton (Kr) ***2:*** 287
Neon (Ne) ***2:*** 355
Radon (Rn) ***3:*** 477
Xenon (Xe) ***3:*** 653

Lanthanides

Cerium (Ce) ***1:*** 113
Dysprosium (Dy) ***1:*** 161
Erbium (Er) ***1:*** 169
Europium (Eu) ***1:*** 175
Gadolinium (Gd) ***2:*** 195
Holmium (Ho) ***2:*** 241
Lutetium (Lu) ***2:*** 313
Neodymium (Nd) ***2:*** 349
Praseodymium (Pr) ***3:*** 453
Promethium (Pm) ***3:*** 459
Samarium (Sm) ***3:*** 505
Terbium (Tb) ***3:*** 585
Thulium (Tm) ***3:*** 605
Ytterbium (Yb) ***3:*** 659

Actinides

Americium (Am) ***1:*** 15
Berkelium (Bk) ***1:*** 49
Californium (Cf) ***1:*** 97
Curium (Cm) ***1:*** 157
Einsteinium (Es) ***1:*** 165
Fermium (Fm) ***1:*** 179
Lawrencium (Lr) ***3:*** 625
Mendelevium (Md) ***3:*** 625
Neptunium (Np) ***2:*** 361
Nobelium (No) ***3:*** 625
Plutonium (Pu) ***3:*** 431
Protactinium (Pa) ***3:*** 465
Thorium (Th) ***3:*** 597
Uranium (U) ***3:*** 639

READER'S GUIDE

Many young people like to play with Lego blocks, tinker-toys, erector sets, and similar building games. It's fun to see how many different ways a few simple shapes can be put together.

The same can be said of chemistry. The world is filled with an untold number of different objects, ranging from crystals and snowflakes to plant and animal cells to plastics and medicines. Yet all of those objects are made from various combinations of only about 100 basic materials: the chemical elements.

Scientists have been intrigued about the idea of an "element" for more than two thousand years. The early Greeks developed complicated schemes that explained everything in nature using only a few basic materials, such as earth, air, fire, and water. The Greeks were wrong in terms of the materials they believed to be "elemental." But they were on the right track in developing the concept that such materials did exist.

By the 1600s, chemists were just beginning to develop a modern definition of an element. An element, they said, was any object that cannot be reduced to some simpler form of matter. Over the next 300 years, research showed that about 100 such materials exist. These materials range from such well known elements as oxygen, hydrogen, iron, gold, and silver to substances that are not at all well known, elements such as neodymium, terbium, rhenium, and seaborgium.

By the mid-1800s, the search for new chemical elements had created a new problem. About 50 elements were known at the time. But no one yet knew how these different elements related to each other, if they did at all. Then, in one of the great coincidences in chemical history, that question was answered independently by two scientists at almost the same time, German chemist Lothar Meyer and Russian chemist Dmitri Mendeleev. (Meyer, however, did not publish his research until 1870, nor did he predict the existence of undiscovered elements as Mendeleev did.)

Meyer and Mendeleev discovered that the elements could be grouped together to make them easier to study. The grouping occurred naturally when the elements were laid out in order, according to their atomic weight. Atomic weight is a quantity indicating atomic mass that tells how much matter there is in an element or how dense it is. The product of Meyer and Mendeleev's research is one of the most famous visual aids in all of science, the periodic table. Nearly every classroom has a copy of this table. It lists all of the known chemical elements, arranged in rows and columns. The elements that lie within a single column or a single row all have characteristics that relate to each other. Chemists and students of chemistry use the periodic table to better understand individual elements and the way the elements are similar to and different from each other.

About *Chemical Elements: From Carbon to Krypton*

Chemical Elements: From Carbon to Krypton is designed as an introduction to the chemical elements. Elements with atomic numbers 1 through 100 are examined in separate entries, while the transfermium elements (elements 101 through 112) are covered in one entry.

Students will find *Chemical Elements* useful in a number of ways. First, it is a valuable source of fundamental information for research reports, science fair projects, classroom demonstrations, and other activities. Second, it can be used to provide more detail about elements and compounds that are only mentioned in other science textbooks or classrooms. Third, it will be an interesting source of information about the building blocks of nature for those who simply want to know more about the elements.

The three-volume set is arranged alphabetically by element name. Each entry contains basic information about the element discussed: its discovery and naming, physical and chemical properties, isotopes, occurrence in nature, methods of extraction, important compounds and uses, and health effects.

The first page of each entry provides basic information about the chemical element: its chemical symbol, atomic number, atomic mass, family, and pronunciation. A diagram of an atom of the element is also shown, with the atom's electrons arranged in various "energy levels" outside the nucleus. Inside the nucleus, the number of protons and neutrons is indicated.

Entries are easy to read and written in a straightforward style. Difficult words are defined within the text. Each entry also includes a "Words to Know" section that defines technical words and scientific terms. This enables students to learn vocabulary appropriate to chemistry without having to consult other sources for definitions.

Added features

Chemical Elements: From Carbon to Krypton includes a number of additional features that help make the connection between elements, minerals, the people who discovered and worked with them, and common uses of the elements.

- Three tables of contents: alphabetically by element name; by atomic number; and by family group provide varied access to the elements.
- A timeline at the beginning of each volume provides a chronology of the discovery of the elements.
- Nearly 200 photographs and illustrations of the elements and products in which they are used bring the elements to life.
- Sidebars provide fascinating supplemental information about scientists, theories, uses of elements, and more.

- Interesting facts about the elements are highlighted in the margins.

- Extensive cross references make it easy to read about related elements. Other elements mentioned within an element's entry are boldfaced upon first mention, serving as a helpful reminder that separate entries are written for these elements.

- A list of sources for further reading for some elements and for general chemistry is found at the end of each volume.

- A comprehensive index quickly points readers to the elements, minerals, and people mentioned in *Chemical Elements: From Carbon to Krypton.*

- A periodic table on the endsheets gives students a quick look at the elements.

Special thanks

The editor wishes to thank imaging guru Randy Bassett for his patience and guidance. Thanks also to Bernard Grunow for his informal assistance in the early stages of the editing phase. Kudos to typesetter Marco Di Vita, who, as always, is in a league by himself. And, finally, a big-time thank-you to soulmate Beth Baker, whose editorial toolbelt, no doubt, needs some duct tape by now.

Comments and suggestions

We welcome your comments on this work as well as suggestions for future science titles. Please write: Editors, *Chemical Elements: From Carbon to Krypton,* U•X•L, 27500 Drake Rd., Farmington Hills, Michigan, 48331-3535; call toll-free: 800-347-4253; send fax to 248-699-8066; or send e-mail via http://www.gale.com.

TIMELINE The Discovery of Elements

Early history	The elements **carbon, sulfur, iron, tin, lead, copper, mercury, silver,** and **gold** are known to humans.
	Pre-a.d. 1600: The elements **arsenic, antimony, bismuth,** and **zinc** are known to humans.
1669	German physician Hennig Brand discovers **phosphorus.**
1735	Swedish chemist Georg Brandt discovers **cobalt.**

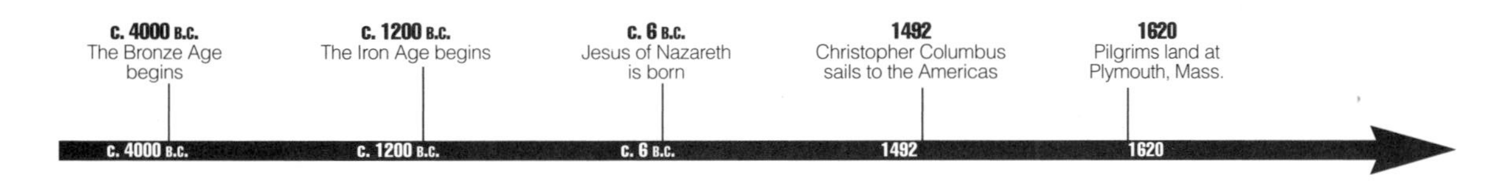

c. 1748	Spanish military leader Don Antonio de Ulloa discovers **platinum.**
1751	Swedish mineralogist Axel Fredrik Cronstedt discovers **nickel.**
1766	English chemist and physicist Henry Cavendish discovers **hydrogen.**
1772	Scottish physician and chemist Daniel Rutherford discovers **nitrogen.**
1774	Swedish chemist Carl Wilhelm Scheele discovers **chlorine.**
1774	Swedish mineralogist Johann Gottlieb Gahn discovers **manganese.**
1774	English chemist Joseph Priestley and Swedish chemist Carl Wilhelm Scheele discover **oxygen.**
1781	Swedish chemist Peter Jacob Hjelm discovers **molybdenum.**
c. 1782	Austrian mineralogist Baron Franz Joseph Müller von Reichenstein discovers **tellurium.**
1783	Spanish scientists Don Fausto D'Elhuyard and Don Juan José D'Elhuyard, and Swedish chemist Carl Wilhelm Scheele discover **tungsten.**

1789	German chemist Martin Klaproth discovers **uranium.**
1789	German chemist Martin Klaproth discovers **zirconium.**
1791	English clergyman William Gregor discovers **titanium.**
1794	Finnish chemist Johan Gadolin discovers **yttrium.**
1797	French chemist Louis-Nicolas Vauquelin discovers **chromium.**
1798	French chemist Louis-Nicolas Vauquelin discovers **beryllium.**
1801	English chemist Charles Hatchett discovers **niobium.**
1801	Spanish-Mexican metallurgist Andrés Manuel del Río discovers **vanadium.**
1802	Swedish chemist and mineralogist Anders Gustaf Ekeberg discovers **tantalum.**
1803	English chemist and physicist William Hyde Wollaston discovers **palladium.**
1803	Swedish chemists Jöns Jakob Berzelius and Wilhelm Hisinger, and German chemist Martin

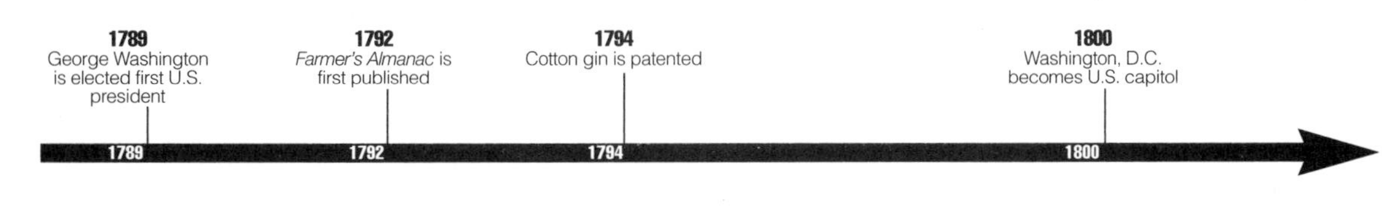

Klaproth discover black rock of Bastnas, Sweden, which led to the discovery of several elements.

1804 English chemist and physicist William Hyde Wollaston discovers **rhodium.**

1804 English chemist Smithson Tennant discovers **osmium.**

1804 English chemist Smithson Tennant discovers **iridium.**

1807 English chemist Sir Humphry Davy discovers **potassium.**

1807 English chemist Sir Humphry Davy discovers **sodium.**

1808 English chemist Sir Humphry Davy discovers **barium.**

1808 English chemist Sir Humphry Davy discovers **strontium.**

1808 English chemist Sir Humphry Davy discovers **calcium.**

1808 English chemist Sir Humphry Davy discovers **magnesium.**

1806
Webster's Dictionary is first published

1804 1806

1808	French chemists Louis Jacques Thênard and Joseph Louis Gay-Lussac discover **boron.**
1811	French chemist Bernard Courtois discovers **iodine.**
1817	Swedish chemist Johan August Arfwedson discovers **lithium.**
1817	German chemist Friedrich Stromeyer discovers **cadmium.**
1818	Swedish chemists Jöns Jakob Berzelius and J. G. Gahn discover **selenium.**
1823	Swedish chemist Jöns Jakob Berzelius discovers **silicon.**
1825	Danish chemist and physicist Hans Christian Oersted discovers **aluminum.**
1826	French chemist Antoine-Jérôme Balard discovers **bromine.**
1828	Swedish chemist Jöns Jakob Berzelius discovers **thorium.**
1830	Swedish chemist Nils Gabriel Sefström rediscovers **vanadium.**
1839	Swedish chemist Carl Gustav Mosander discovers **cerium.**

1839	Swedish chemist Carl Gustav Mosander discovers **lanthanum.**
1843	Swedish chemist Carl Gustav Mosander discovers **terbium.**
1843	Swedish chemist Carl Gustav Mosander discovers **erbium.**
1844	Russian chemist Carl Ernst Claus discovers **ruthenium.**
c. 1861	German chemists Robert Bunsen and Gustav Kirchhoff discovers **cesium.**
c. 1861	German chemists Robert Bunsen and Gustav Kirchhoff discovers **rubidium.**
1861	British physicist Sir William Crookes discovers **thallium.**
1863	German chemists Ferdinand Reich and Hieronymus Theodor Richter discovers **indium.**
1875	Paul-émile Lecoq de Boisbaudran discovers **gallium.**
1878	Jean-Charles-Galissard de Marignac receives partial credit for the discovery of **ytterbium.**
1879	Swedish chemist Per Teodor Cleve discovers **holmium.**

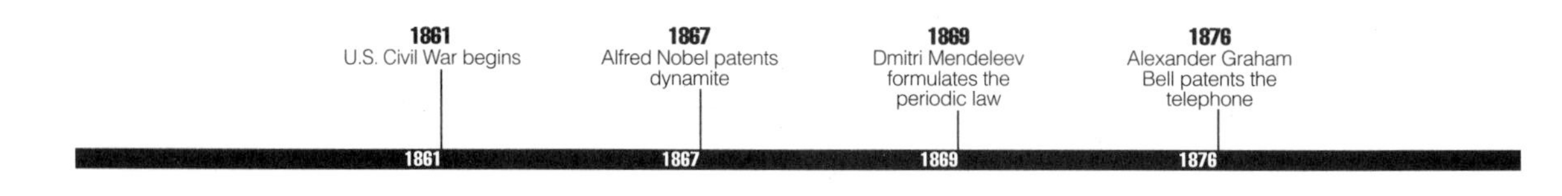

1879	Swedish chemist Per Teodor Cleve discovers **thulium.**
1879	Swedish chemist Lars Nilson discovers **scandium.**
1879	Swedish chemist Lars Nilson receives partial credit for the discovery of **ytterbium.**
1880	French chemist Paul-Émile Lecoq de Boisbaudran discovers **samarium.**
1880	French chemist Jean-Charles-Galissard de Marignac discovers **gadolinium.**
1885	Austrian chemist Carl Auer (Baron von Welsbach) discovers **praseodymium.**
1885	Austrian chemist Carl Auer (Baron von Welsbach) discovers **neodymium.**
1885	German chemist Clemens Alexander Winkler discovers **germanium.**
1886	French chemist Henri Moissan discovers **fluorine.**
1886	French chemist Paul-Émile Lecoq de Boisbaudran discovers **dysprosium.**
1894	English chemists Lord Rayleigh and William Ramsay discover **argon.**

1884
A worldwide system of standard time is adopted

1888
George Eastman introduces the Kodak camera

1892
Rudolf Diesel patents the internal-combustion engine

1884 1888 1892

Timeline: The Discovery of Elements

1895	English chemist Sir William Ramsay and Swedish chemists Per Teodor Cleve and Nils Abraham Langlet discover **helium.**
1898	English chemists William Ramsay and Morris Travers discover **krypton.**
1898	English chemists William Ramsay and Morris Travers discover **neon.**
1898	English chemists William Ramsay and Morris Travers discover **xenon.**
1898	French physicists Marie and Pierre Curie discover **polonium.**
1898	French physicists Marie and Pierre Curie discover **radium.**
1899	French chemist André Debierne discovers **actinium.**
1900	German physicist Friedrich Ernst Dorn discovers **radon.**
1901	French chemist Eugène-Anatole Demarçay discovers **europium.**
1907	French chemist Georges Urbain discovers **lutetium.**

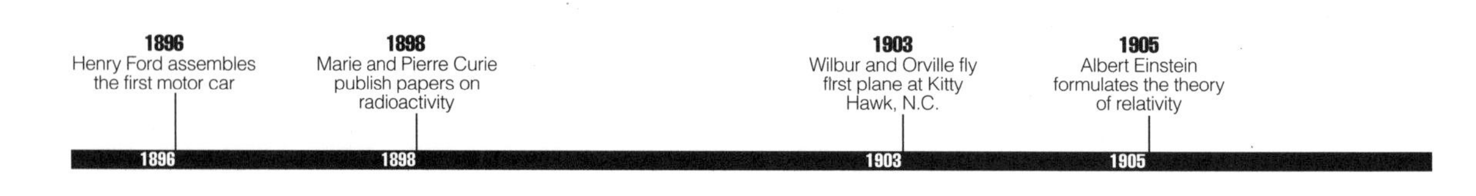

1907	French chemist Georges Urbain receives partial credit for the discovery of **ytterbium.**
1917	German physicists Lise Meitner and Otto Hahn discover **protactinium.**
1923	Dutch physicist Dirk Coster and Hungarian chemist George Charles de Hevesy discover **hafnium.**
1925	German chemists Walter Noddack, Ida Tacke, and Otto Berg discover **rhenium.**
1933	French chemist Marguerite Perey discovers **francium.**
1939	Italian physicist Emilio Segrè and his colleague Carlo Perrier discover **technetium.**
1940	Edwin M. McMillan (1907–91) and Philip H. Abelson prepare **neptunium.**
1940	Dale R. Corson, Kenneth R. Mackenzie, and Emilio Segrè discover **astatine.**
1940	University of California at Berkeley researcher Glenn Seaborg and others prepare **plutonium.**
1944	University of California at Berkeley researchers Glenn Seaborg, Albert Ghiorso, Ralph A.

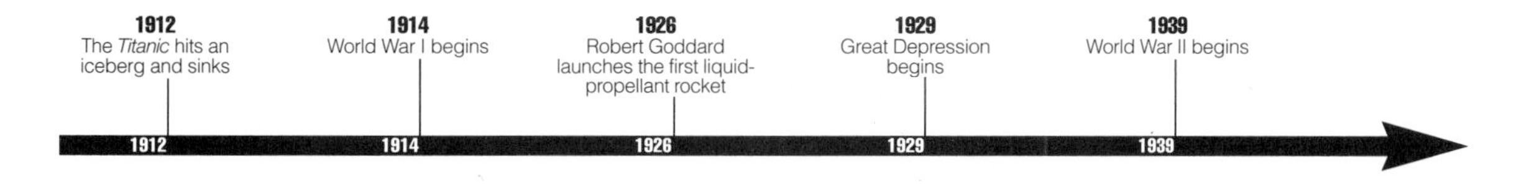

	James, and Leon O. Morgan prepare **americium.**
1944	University of California at Berkeley researchers Glenn Seaborg, Albert Ghiorso, and Ralph A. James prepare **curium.**
1945	Scientists at the Oak Ridge Laboratory in Oak Ridge, Tennessee, discover **promethium.**
1949	University of California at Berkeley researchers prepare **berkelium.**
1950	University of California at Berkeley researchers Glenn Seaborg, Albert Ghiorso, Kenneth Street, Jr., and Stanley G. Thompson prepare **californium.**
1954	University of California at Berkeley researchers prepare **einsteinium.**
1954	University of California at Berkeley researcher Albert Ghiorso and others prepare **fermium.**
1960s & 1970s	Researchers at the Joint Institute of Nuclear Research, in Dubna, Russia; the Lawrence Berkeley Laboratory at the University of California at Berkeley; and the Institute for Heavy Ion Research in Darmstadt, Germany, continue to prepare new transfermium elements.

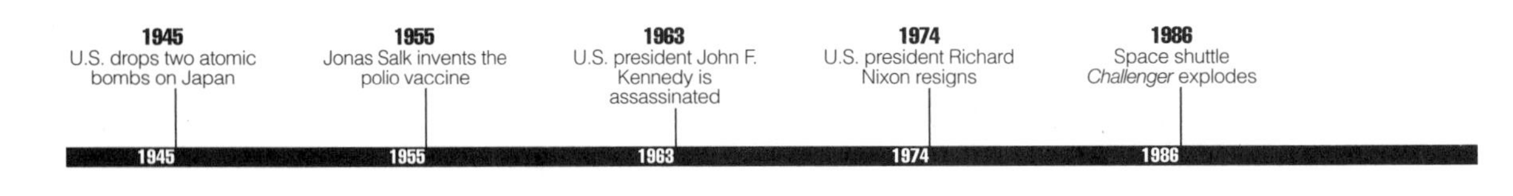

WORDS TO KNOW

A

Abrasive a powdery material used to grind or polish other materials

Absolute zero the lowest temperature possible, about –273°C (–459°F)

Actinide family elements in the periodic table with atomic numbers 90 through 103

Alchemy a kind of pre-science that existed from about 500 B.C. to about the end of the 16th century

Alkali metal an element in Group 1 (IA) of the periodic table

Alkali a chemical with properties opposite those of an acid

Alkaline earth metal an element found in Group 2 (IIA) of the periodic table

Allotropes forms of an element with different physical and chemical properties

Alloy a mixture of two or more metals that has properties different from those of the individual metals

Alpha particles tiny, atom-sized particles that can destroy cells

Alpha radiation a form of radiation that consists of very fast moving alpha particles and helium atoms without their electrons

Amalgam a combination of mercury and at least one other metal

Amorphous without crystalline shape

Anhydrous ammonia dry ammonia gas

Antiseptic a chemical that stops the growth of germs

Aqua regia a mixture of hydrochloric and nitric acids that often reacts with materials that do not react with either acid separately

B

Battery a device for changing chemical energy into electrical energy

Biochemistry the field of chemistry concerned with the study of compounds found in living organisms

Biocompatible not causing a reaction when placed into the body

Bipolar disorder a condition in which a person experiences wild mood swings

Brass an alloy of copper and zinc

Bronze Age a period in human history ranging from about 3500 B.C. to 1000 B.C., when bronze was widely used for weapons, utensils, and ornamental objects

Bronze an alloy of copper and tin

Buckminsterfullerene full name for buckyball or fullerene; *see* Buckyball

Buckyball an allotrope of carbon whose 60 carbon atoms are arranged in a sphere-like form

C

Capacitor an electrical device, somewhat like a battery, that collects and then stores up electrical charges

Carat a unit of weight for gold and other precious metals, equal to one fifth of a gram, or 200 milligrams

Carbon arc lamp a lamp for producing very bright white light

Carbon-14 dating a technique that allows archaeologists to estimate the age of once-living materials by using the knowledge that carbon-14 is found in all living carbon materials; once an organism dies, no more carbon-14 remains

Cassiterite an ore of tin containing tin oxide, the major commercial source of tin metal

Catalyst a substance used to speed up or slow down a chemical reaction without undergoing any change itself

Chalcogens elements in Group 16 (VIA) of the periodic table

Chemical reagent a substance, such as an acid or an alkali, used to study other substances

Chlorofluorocarbons (CFCs) a family of chemical compounds consisting of carbon, fluorine, and chlorine that were once used widely as propellants in commercial sprays but regulated in the United States since 1987 because of their harmful environmental effects

Corrosive agent a material that tends to vigorously react or eat away at something

Cyclotron a particle accelerator, or "atom smasher," in which small particles, such as protons, are made to travel very fast and then collide with atoms, causing the atoms to break apart

D

Density the mass of a substance per unit volume

Diagnosis finding out what medical problems a person may have

Distillation a process by which two or more liquids can be separated from each other by heating them to their boiling points

"Doped" containing a small amount of a material as an impurity

Ductile capable of being drawn into thin wires

E

Earth in mineralogy, a naturally occurring form of an element, often an oxide of the element

Electrolysis a process by which a compound is broken down by passing an electric current through it

Electroplating the process by which a thin layer of one metal is laid down on top of a second metal

Enzyme a substance that stimulates certain chemical reactions in the body

F

Fabrication shaping, molding, bending, cutting, and working with a metal

Fission the process by which large atoms break apart, releasing large amounts of energy, smaller atoms, and neutrons in the process

Fly ash the powdery material produced during the production of iron or some other metal

Frasch method a method for removing sulfur from underground mines by pumping hot air and water down a set of pipes

Fuel cell any system that uses chemical reactions to produce electricity

Fullerene alternative name for buckyball; *see* Buckyball

G

Galvanizing the process of laying down a thin layer of zinc on the surface of a second metal

Gamma rays a form of radiation similar to X rays

H

Half life the time it takes for half of a sample of a radioactive element to break down

Halogen one of the elements in Group 17 (VIIA) of the periodic table

Heat exchange medium a material that picks up heat in one place and carries it to another place

Hydrocarbons compounds made of carbon and hydrogen

Hypoallergenic not causing an allergic reaction

I

Inactive does not react with any other element

Inert gases *see* **Noble gases**

Inert not very active

Isotope two or more forms of an element that differ from each other according to their mass number

L

Lanthanide family the elements in the periodic table with atomic numbers 58 through 71

Laser a device for making very intense light of one very specific color that is intensified many times over

Liquid air air that has been cooled to a very low temperature

Luminescence the property of giving off light without giving off heat

M

Machining the bending, cutting, and shaping of a metal by mechanical means

"Magic number" the number of protons and/or neutrons in an atom that tend to make the atom stable (not radioactive)

Magnetic field the space around an electric current or a magnet in which a magnetic force can be observed

Malleable capable of being hammered into thin sheets

Metals elements that have a shiny surface, are good conductors of heat and electricity, and can be melted, hammered into thin sheets, and drawn into thin wires

Metalloid an element that has characteristics of both metals and non-metals

Metallurgy the art and science of working with metals

Micronutrient a substance needed in very small amounts to maintain good health

Misch metal a metal that contains different rare earth elements and has the unusual property of giving off a spark when struck

Mohs scale a way of expressing the hardness of a material

Mordant a material that helps a dye stick to cloth

N

Nanotubes long, thin, and extremely tiny tubes

Native not combined with any other element

Neutron radiography a technique that uses neutrons to study the internal composition of material

Nickel allergy a health condition caused by exposure to nickel metal

Nitrogen fixation the process of converting nitrogen as an element to a compound that contains nitrogen

Noble gases elements in Group 18 (VIIIA) of the periodic table

Noble metals see **Platinum family**

Non-metals elements that do not have the properties of metals

Nuclear fission a process in which neutrons collide with the nucleus of a plutonium or uranium atom, causing it to split apart with the release of very large amounts of energy

Nuclear reactor a device in which nuclear reactions occur

O

Optical fiber a thin strand of glass through which light passes; the light carries a message, much as an electric current carries a message through a telephone wire

Ore a mineral compound that is mined for one of the elements it contains, usually a metal element

Organic chemistry the study of the carbon compounds

Oxidizing agent a chemical substance that gives up or takes on electrons from another substance

Ozone a form of oxygen that filters out harmful radiation from the sun

P

Particle accelerator ("atom smasher") a device used to cause small particles, such as protons, to move at very high speeds

Periodic law a way of organizing the chemical elements to show how they are related to each other

Periodic table a chart that shows how chemical elements are related to each other

Phosphor a material that gives off light when struck by electrons

Photosynthesis the process by which plants convert carbon dioxide and water to carbohydrates (starches and sugars)

Platinum family a group of elements that occur close to platinum in the periodic table and with platinum in the Earth's surface

Polymerization the process by which many thousands of individual tetrafluoroethlylene (TFE) molecules join together to make one very large molecule

Potash a potassium compound that forms when wood burns

Precious metal a metal that is rare, desirable, and, therefore, expensive

Proteins compounds that are vital to the building and growth of cells

Pyrophoric gives off sparks when scratched

Q

Quarry a large hole in the ground from which useful minerals are taken

R

Radiation energy transmitted in the form of electromagnetic waves or subatomic particles

Radioactive isotope an isotope that breaks apart and gives off some form of radiation

Radioactive tracer an isotope whose movement in the body can be followed because of the radiation it gives off

Radioactivity the process by which an isotope or element breaks down and gives off some form of radiation

Rare earth elements *see* **Lanthanide family**

Reactive combines with other substances relatively easily

Refractory a material that can withstand very high temperatures and reflects heat back away from itself

Rodenticide a poison used to kill rats and mice

Rusting a process by which a metal combines with oxygen

S

Salt dome a large mass of salt found underground

Semiconductor a material that conducts an electric current, but not nearly as well as metals

Silver plating a process by which a very thin layer of silver metal is laid down on top of another metal

Slag a mixture of materials that separates from a metal during its purification and floats on top of the molten metal

Slurry a soup-like mixture of crushed ore and water

Solder an alloy that can be melted and then used to join two metals to each other

Spectra the lines produced when chemical elements are heated

Spectroscope A device for analyzing the light produced when an element is heated

Spectroscopy the process of analyzing light produced when an element is heated

Spectrum (plural: spectra) the pattern of light given off by a glowing object, such as a star

Stable not likely to react with other materials

Sublimation the process by which a solid changes directly to a gas when heated, without first changing to a liquid

Superalloy an alloy made of iron, cobalt, or nickel that has special properties, such as the ability to withstand high temperatures and attack by oxygen

Superconductivity the tendency of an electric current to flow through a material without resistance

Superconductor a material that has no resistance to the flow of electricity; once an electrical current begins flowing in the material, it continues to flow forever

Superheated water water that is hotter than its boiling point, but which has not started to boil

Surface tension a property of liquids that makes them act like they are covered with a skin

T

Tarnishing oxidizing; reacting with oxygen in the air

Tensile capable of being stretched without breaking

Thermocouple a device for measuring very high temperatures

Tin cry a screeching-like sound made when tin metal is bent

Tin disease a change that takes place in materials containing tin when the material is cooled to temperatures below 13°C for long periods of time, when solid tin turns to a crumbly powder

Tincture a solution made by dissolving a substance in alcohol

Tinplate a type of metal consisting of thin protective coating of tin deposited on the outer surface of some other metal

Toxic poisonous

Trace element an element that is needed in very small amounts for the proper growth of a plant or animal

Tracer a radioactive isotope whose presence in a system can easily be detected

Transfermium element any element with an atomic number greater than 100

Transistor a device used to control the flow of electricity in a circuit

Transition metal an element in Groups 3 through 12 of the periodic table

Transuranium element an element with an atomic number greater than 92

U

Ultraviolet (UV) radiation electromagnetic radiation (energy) of a wavelength just shorter than the violet (shortest

wavelength) end of the visible light spectrum and thus with higher energy than visible light

V

Vulcanizing the process by which soft rubber is converted to a harder, longer-lasting product

W

Workability the ability to work with a metal to get it into a desired shape or thickness

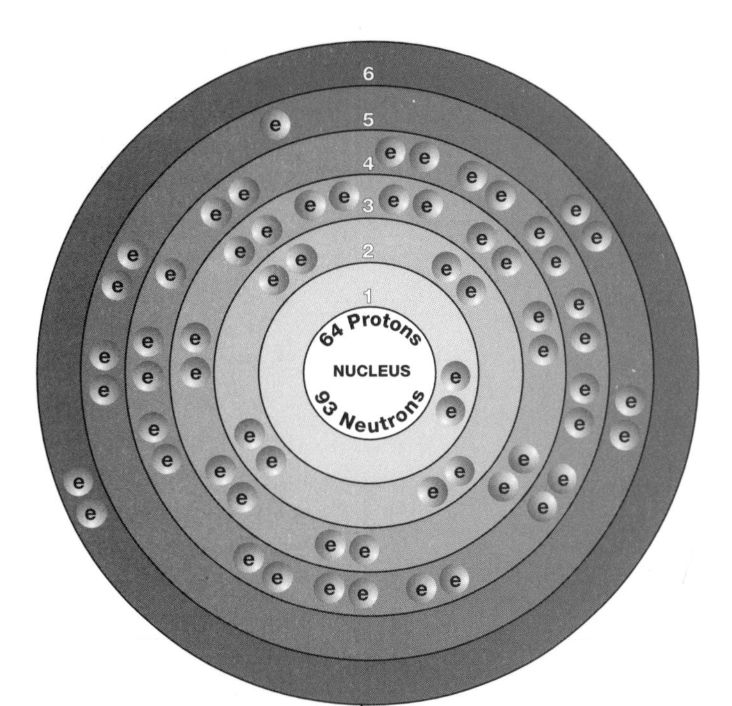

GADOLINIUM

Overview

Gadolinium was named for Finnish chemist Johan Gadolin (1760–1852). Gadolin served for many years as professor of chemistry at the University of Åbo in Finland. He was the first person to study an unusual black stone discovered near the town of Ytterby, Sweden, in 1787. The stone was an unusually important discovery. Chemists isolated nine new elements from the stone, one of which was gadolinium.

Gadolinium is in Row 6 of the periodic table. The periodic table is a chart that shows how chemical elements are related to each other. The elements in Row 6 are called rare earth elements. They really aren't rare, but they *are* difficult to separate. For this reason, scientists know less about the rare earth elements than they do about most other elements. The rare earth metals are also called lanthanides. That name comes from the first element in Row 6, **lanthanum.**

SYMBOL
Gd

ATOMIC NUMBER
64

ATOMIC MASS
157.25

FAMILY
Lanthanide
(rare earth metal)

PRONUNCIATION
gad-uh-LIN-ee-um

Discovery and naming

Two unusual rocks were discovered in Sweden near the end of the eighteenth century. The rocks were unusual because they both contained a complex mixture of substances. Chemists worked for nearly a century to separate the mixtures and find

WORDS TO KNOW

Ductile capable of being drawn into a thin wire

Isotopes two or more forms of an element that differ from each other according to their mass number

Lanthanides the elements in the periodic table with atomic numbers 58 through 71

Malleable capable of being hammered into thin sheets

Neutron radiography a technique that uses neutrons to study the internal composition of material

Nuclear fission the process in which large atoms break apart, releasing large amounts of energy, smaller atoms, and neutrons in the process

Phosphor a material that gives off light when struck by electrons

Radioactive isotope an isotope that breaks apart and gives off some form of radiation

Rare earth elements *see* **Lanthanides**

Toxic poisonous

out what they were. All fifteen rare earth elements were first discovered in the two Swedish rocks.

One rock contained a mineral that had never been seen before, cerite. Cerite was first discovered in 1803. The last new element found in cerite was not identified until almost a century later, in 1901. In 1880, French chemist Jean-Charles Galissard de Marignac (1817–94) was studying a new material found in cerite called samaria. Earlier chemists had identified samaria as a new element.

Marignac found that samaria was not a pure element. Instead, it consisted of two parts, which he called samaria and gadolinia. He believed each was a new element. He was right about gadolinia, but wrong about samaria.

Physical properties

Gadolinium has a shiny metallic luster with a slight yellowish tint. It is both ductile and malleable. Ductile means capable of being made into wires. Malleable means capable of being hammered or rolled into thin sheets. It has a melting point of 1,312°C (2,394°F) and a boiling point of about 3,000°C (5,400°F). Its density is 7.87 grams per cubic centimeter.

Few elements are as strongly magnetic as gadolinium. It also has the highest neutron-absorbing ability of any element. A piece of gadolinium stops neutrons better than any other element.

Chemical properties

Gadolinium metal is not especially reactive. It dissolves in acids and reacts slowly with cold water. It also reacts with **oxygen** at high temperatures.

Occurrence in nature

The abundance of gadolinium in the Earth's surface is estimated at about 4.5 to 6.4 parts per million. That would make it one of the most abundant of the rare earth elements. It ranks above **bromine** and **uranium,** but just below **lead** and **boron** in order of abundance. Some minerals in which it occurs are monazite, bastnasite, samarskite, gadolinite, and xenotime.

Isotopes

Seven naturally occurring isotopes of gadolinium are known. They are gadolinium-152, gadolinium-154, gadolinium-155, gadolinium-156, gadolinium-157, gadolinium-158, and gadolinium-160. Isotopes are two or more forms of an element. Isotopes differ from each other according to their mass number. The number written to the right of the element's name is the mass number. The mass number represents the number of protons plus neutrons in the nucleus of an atom of the element. The number of protons determines the element, but the number of neutrons in the atom of any one element can vary. Each variation is an isotope. One of gadolinium's naturally occurring isotopes is radioactive—gadolinium-152. A radioactive isotope is one that breaks apart and gives off some form of radiation.

Nine new elements, including gadolinium, came from an unusual black stone found in Ytterby, Sweden, in 1787.

Radioactive isotopes can also be produced artificially when very small particles are fired at atoms. These particles stick in the atoms and make them radioactive. At least eleven artificial radioactive isotopes have been produced. Some of these are used in medicine. For example, gadolinium-153 is used to study the composition of bones. The radiation it gives off acts like X rays, penetrating the bones to reveal the minerals present.

Gadolinium is also used in another X-ray-like technique called neutron radiography. In this technique, neutrons are fired through a sample of material. The neutrons act somewhat like X rays. They show the structure of the material. Adding gadolinium to the back side of the material makes the neutron image easier to read. Neutron radiography is especially useful because one can look for damage inside a piece of metal without having to take the material apart.

Extraction

The method for obtaining gadolinium from its ores is similar to that for other rare earth elements. The ore is converted into gadolinium chloride ($GdCl_3$) or gadolinium fluoride (GdF_3). Passing an electric current through the first compound releases pure gadolinium:

$$2GdCl_3 \xrightarrow{\text{electric current}} 2Gd + 3Cl_2$$

Adding calcium to the second compound also releases pure gadolinium:

$$3Ca + 2GdF_3 \rightarrow 3CaF_2 + 2Gd$$

Special gadolinium minerals called yttrium garnets are used in microwave ovens.

Uses and compounds

Gadolinium is used in control rods in nuclear power plants. Energy produced during nuclear fission is used to generate electricity. Nuclear fission is the process in which large atoms (usually uranium or **plutonium**) break apart, releasing energy. The smaller atoms produced are called fission products and are radioactive.

Neutrons are also produced in the reaction. In order for a nuclear power plant to work properly, the number of neutrons must be carefully controlled. Rods containing gadolinium are raised out of or lowered into the reactor. This allows more or fewer neutrons to remain in the reaction.

Gadolinium is used to locate tumors in the inner ear.

Gadolinium also has medical uses. It is used to locate the presence of tumors in the inner ear. Gadolinium is injected into the blood stream. It then goes to any tumor that happens to be present in the ear. The tumor appears darker when seen with X rays.

Gadolinium compounds are used as phosphors in television tubes. A phosphor is a material that shines when struck by electrons. The color of the phosphor depends on the elements of which it is made.

What on Earth is an "Earth"?

Books on the chemical elements sometimes talk about gadolin*ium* and gadolin*ia*; about erb*ium* and sometimes erb*ia*; sometimes about samar*ium* and sometimes samar*ia*. Is the *-ium* ending any different from the *-ia* ending?

The answer is yes. Almost all metal names end in *-ium* or just *-um*, like sod*ium*, potass*ium*, magnes*ium*, alumin*um*, and gadolin*ium*. The ending *-ia* or just *-a*, on the other hand, stands for the form in which an element occurs in the earth. When miners take gadolinium out of the earth, it is called *gadolinia*.

The natural form of the element is often called an "earth." Gadolinium is the element that comes from the earth, gadolinia. Earths are compounds of the element and one or more other element. Two common combining elements are oxygen and sulfur. **For example, gadolinia contains gadolinium oxide (Gd_2O_3).**

These terms can be confusing when reading the history of chemical elements. Many elements were first discovered not in their pure form, but as compounds—as earths.

Gadolinium is also found in alloys and special minerals known as yttrium garnets. An alloy is made by melting and mixing two or more metals. The mixture has properties different from those of the individual metals. Gadolinium alloys are easier to work with than alloys without gadolinium. Gadolinium yttrium garnets are used in microwave ovens to produce the microwaves.

Health effects

Not many details of the health effects of gadolinium are known. It is usually handled as if it were very toxic.

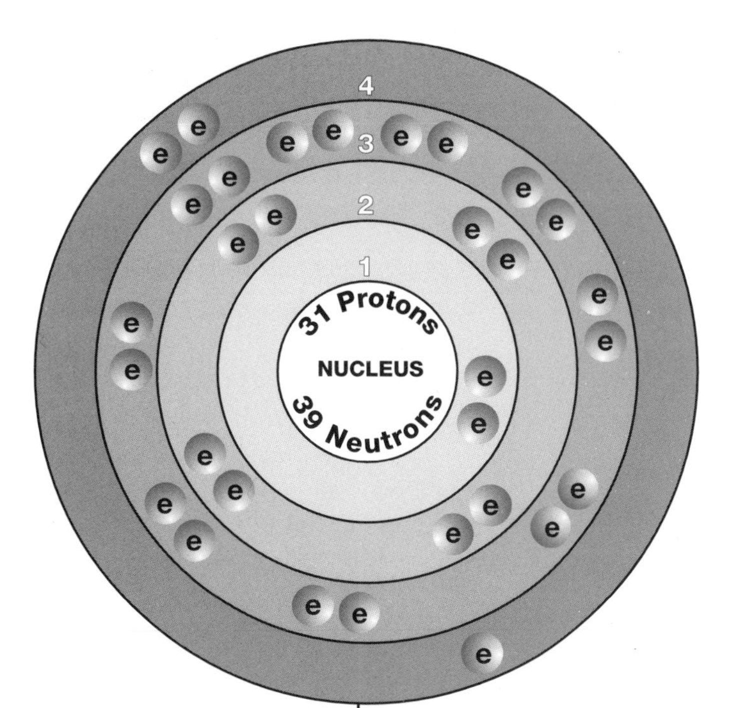

GALLIUM

SYMBOL
Ga

ATOMIC NUMBER
31

ATOMIC MASS
69.72

FAMILY
Group 13 (IIIA)
Aluminum

PRONUNCIATION
GA-le-um

Overview

In the late 1860s Russian chemist Dmitri Mendeleev (1834–1907) made one of the greatest discoveries in modern chemistry: the periodic law. The periodic law describes how chemical elements are related to each other. These elements are in the periodic table. This is a chart that lists all of the chemical elements and sorts them into groups based on similarities. Elements in vertical columns are similar to each other in many ways.

When Mendeleev first proposed the periodic law, he made a troubling discovery. There were a few empty spots in his table. For example, the box set aside for element number 31 was empty. No element had been found that belonged in that box.

Part of Mendeleev's genius was what he did next. He said that an "element number 31" did exist. Scientists simply had not found it yet. But Mendeleev described what the element would be like. He based his prediction on elements on all sides of the box for element number 31. He said it would be similar to **aluminum** (in box 13, above 31) and **indium** (in box 49, below 31). He named this missing element eka-aluminum.

Using Mendeleev's periodic law, the element was soon found. It was discovered by French chemist Paul Émile Lecoq de Boisbaudran in 1875.

Until recently, gallium had few applications. Then, some of its compounds were discovered to have unusual properties when exposed to light. These properties make gallium an important and essential element in many electronic devices.

Discovery and naming

Lecoq de Boisbaudran did not discover gallium by accident. For 15 years, he had been studying the spectra of the chemical elements. Spectra (singular: spectrum) are the lines produced when chemical elements are heated. Each element produces its own distinctive set of lines, or spectra. An element can be identified in a sample by the spectrum it produces.

Lecoq de Boisbaudran knew that the element between aluminum and indium was missing. He also knew about Mendeleev's prediction. Lecoq de Boisbaudran wanted to learn more about the spectra of elements. He thought that element number 31 might be found in **zinc** ores. Zinc has an atomic number of 30, so it is next to gallium on the periodic table.

Lecoq de Boisbaudran had to work through a large amount of zinc ore. But his hunch turned out to be correct. The missing element was present in the ore, but only in very small amounts. Finally, in August 1875, Lecoq de Boisbaudran reported that "the new substance gave under the action of the electric spark a spectrum composed chiefly of a violet ray, narrow, readily visible, and [located at] about 417 on the scale of wave lengths."

Later in the same year, Lecoq de Boisbaudran isolated gallium metal. He was given several tons of zinc ore by miners for his research. Out of this ore, he was able to produce a few grams of nearly pure gallium.

Lecoq de Boisbaudran proposed the name gallium for the new element. The name was given in honor of the ancient name for France, Gallia.

Physical properties

Gallium is a soft, silvery metal with a shiny surface. In some ways, however, it is very un-metal-like. It is so soft that it can

WORDS TO KNOW

Ductile capable of being drawn into thin wires

Isotopes two or more forms of an element that differ from each other according to their mass number

Laser a device for making very intense light of one very specific color that is intensified many times over

Periodic table a chart that shows how chemical elements are related to each other

Radioactive isotope an isotope that breaks apart and gives off some form of radiation

Semiconductor a material that conducts an electrical current

Spectra the lines produced when chemical elements are heated

Transistor a device used to control the flow of electricity in a circuit

Gallium melts when held in the hand.

be cut with a knife. It has a very low melting point of only 29.7°C (85.5°F). A sample of gallium will melt if held in the human hand (body temperature, about 37°C).

Another unusual property is that gallium can be supercooled rather easily. Supercooling is the cooling of a substance below its freezing point without it becoming a solid. Gallium is a liquid at 30°C, so one would expect it to become a solid at 29.7°C. Instead it is fairly easy to cool gallium to below 29.7°C without having it solidify.

Gallium's boiling point is about 2,400°C (4,400°F) and its density is 5.9037 grams per cubic centimeter.

Chemical properties

Gallium is a fairly reactive element. It combines with most non-metals at high temperatures, and it reacts with both acids and alkalis. An alkali is a chemical with properties opposite

those of an acid. Sodium hydroxide (common lye, such as Drano) and bleach are examples of alkalis.

Occurrence in nature

Gallium is a moderately abundant element in the Earth's crust. Its abundance has been estimated to be about 5 parts per million. It is found primarily in combination with zinc and aluminum ores. It is also found in germanite, an ore of **copper** sulfide (CuS).

The United States produces no gallium. The largest producers are Australia, Russia, France, and Germany.

Gallium-67 has been used to look for cancer in the liver, spleen, bowels, breasts, thymus, kidneys, and bones.

Isotopes

Two naturally occurring isotopes of gallium are known: gallium-69 and gallium-71. Isotopes are two or more forms of an element. Isotopes differ from each other according to their mass number. The number written to the right of the element's name is the mass number. The mass number represents the number of protons plus neutrons in the nucleus of an atom of the element. The number of protons determines the element, but the number of neutrons in the atom of any one element can vary. Each variation is an isotope.

About a dozen radioactive isotopes of gallium are known also. A radioactive isotope is one that breaks apart and gives off some form of radiation. Radioactive isotopes are produced when very small particles are fired at atoms. These particles stick in the atoms and make them radioactive.

One radioactive isotope of gallium, gallium-67, has long been used in medicine. This isotope has a tendency to seek out cancer cells in the body. Its presence in a cell can be detected by the radiation it gives off. By giving a patient a dose of gallium-67, a doctor can determine whether the patient has cancer. Gallium-67 has been used to look for cancer in the liver, spleen, bowels, breasts, thymus, kidneys, and bones.

Extraction

Pure gallium metal can be prepared by passing an electric current through a gallium compound, such as gallium oxide (Ga_2O_3):

$$2Ga_2O_3 \xrightarrow{\text{electric current}} 4Ga + 3O_2$$

Close-up view of LED numerical display. Gallium arsenide is used to make LEDs.

Uses and compounds

About 95 percent of all gallium produced is used to make a single compound, gallium arsenide (GaAs). Gallium arsenide has the ability to convert an electrical current directly into light. The lighted numbers on hand-held calculators, for example, are produced by a device known as a light-emitting diode (LED). Gallium arsenide is used to make LEDs. An LED allows an electric current to flow in one side, but not the other. When it flows into a piece of gallium arsenide, a flash of light is produced. When a button is pushed on a calculator,

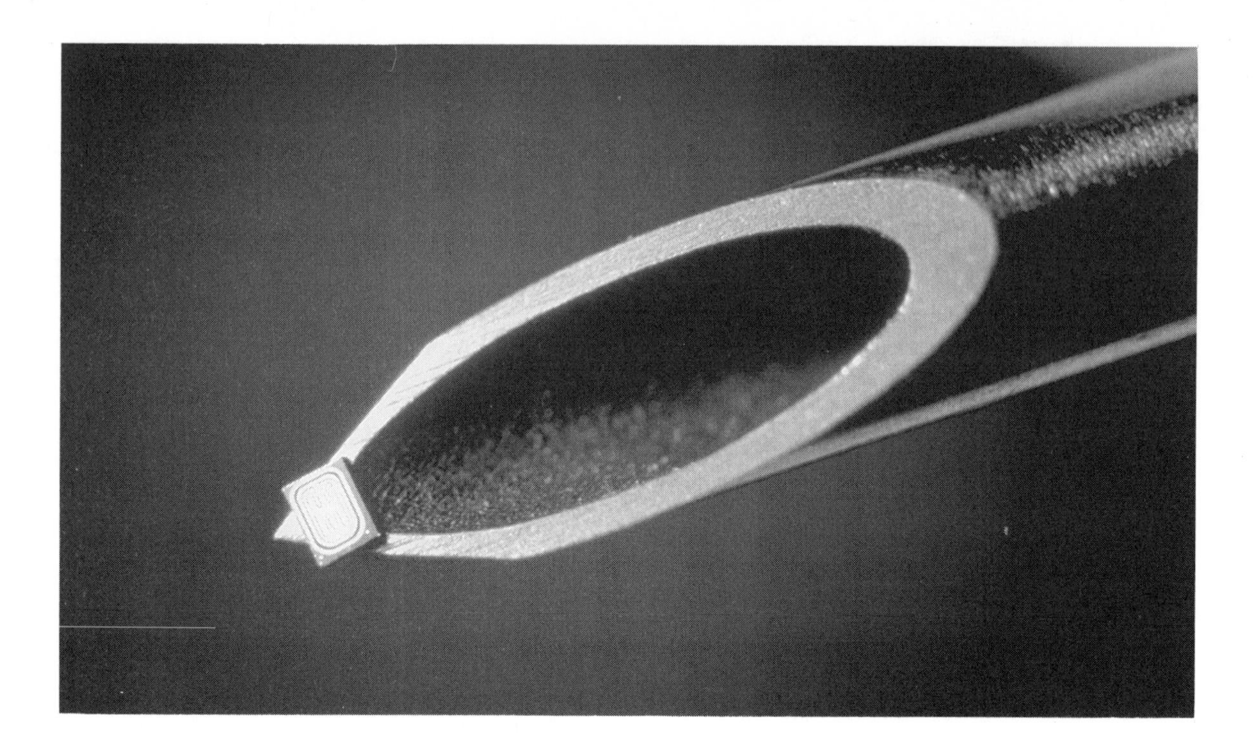

Gallium arsenide is used in the production of transistors.

a circuit is closed. The electric current flows into an LED and produces a light.

Similar devices are used in making lasers. An electric current passes into a piece of gallium arsenide. The current produces an intense beam of laser light. A laser is a device for producing very bright light of a single color. Gallium arsenide lasers are used in a number of applications. The laser that operates a compact disc (CD) player, for example, may contain a piece of gallium arsenide.

Gallium arsenide is also used to make transistors. A transistor is a device used to control the flow of electricity in a circuit. Gallium arsenide has many of the properties of a semiconductor. A semiconductor is a material that conducts an electrical current, but not as well as a metal, such as **silver** or copper. Gallium arsenide has one big advantage over **silicon,** another element used in transistors. Gallium arsenide produces less heat. Therefore, more transistors can work together at the same time to produce a higher computing capacity.

Gallium arsenide is also used in photovoltaic cells. These devices turn sunlight into electricity. Many people believe that photovoltaic cells will someday replace coal-fired generating and nuclear power plants as the major source of electricity.

Health effects

Gallium and its compounds are somewhat hazardous to the health of humans and animals. They produce a metallic taste in the mouth, skin rash, and a decrease in the production of blood cells. Gallium and its compounds should be handled with caution.

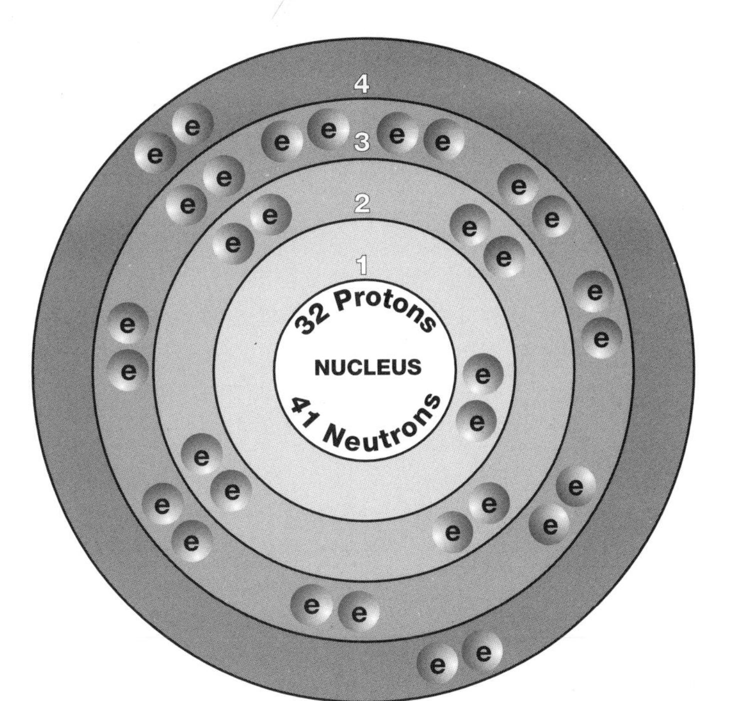

GERMANIUM

Overview

Germanium is a metalloid. A metalloid is an element that has characteristics of both metals and non-metals. Germanium is located in the middle of the carbon family, which is Group 14 (IVA) in the periodic table. The periodic table is a chart that shows how chemical elements are related to each other. **Carbon** and **silicon** are above germanium and **tin** and **lead** are below it.

The existence of germanium was predicted by Russian chemist Dmitri Mendeleev (1834–1907), who developed the periodic table. Mendeleev predicted a new element would be found to fill an empty spot on the table. He was proved correct in 1886.

Until the 1950s, there were no important uses for germanium. Then, the transistor was invented. A transistor is a device used to control the flow of electricity in a circuit. Today, germanium is used in making a number of electronic devices, including the transistor.

Discovery and naming

In the 1860s, Mendeleev wondered if the chemical elements could be arranged in any systematic way. Are all chemical ele-

SYMBOL
Ge

ATOMIC NUMBER
32

ATOMIC MASS
72.59

FAMILY
Group 14 (IVA)
Carbon

PRONUNCIATION
jur-MAY-nee-um

ments very different from each other, he asked. Do they have certain common properties?

He explored a number of ways of arranging the elements. Finally, he decided to arrange them according to their atomic weights. He found that doing so resulted in a pattern. After awhile, each element could be placed in a position beneath one or more elements before it. Mendeleev described his result in the periodic law. The periodic table is the most common way of illustrating the periodic law. Elements that are similar to each other fall into the same group. For example, the elements in Group 1 (IA) are like each other in many ways.

Mendeleev found that his periodic table made sense, however, only if he left some blank spaces in it. His table had a blank space for element number 32. No element existed that had properties like silicon (number 14) and could be put beneath it in the periodic table.

This finding did not disturb Mendeleev. Element number 32 simply had not been discovered yet, he said.

A number of chemists took up Mendeleev's challenge. In 1885, a new ore was discovered in a mine near Freiberg, Germany. **Silver** and sulfur were found in the ore, but about seven percent of the ore could not be identified. The ore was sent to German chemist Clemens Alexander Winkler (1838–1904). At the time, Winkler was professor of chemical technology and analytical chemistry at the Freiberg School of Mines. He was convinced that the new ore contained a new element. He isolated the new element from the ore and named it germanium, in honor of Winkler's native country, Germany.

Winkler discovered that the properties of germanium were very similar to those that Mendeleev had predicted 15 years earlier. For example, Mendeleev thought element 32 would have a density of 5.5 grams per cubic centimeter. The actual density for germanium is 5.47 grams per cubic centimeter. Mendeleev had based his predictions on the new element's place in the periodic table. His success in making these predictions gave chemists a great deal of confidence in the periodic table. They came to see how useful it could be in their research.

WORDS TO KNOW

Catalyst a substance used to speed up or slow down a chemical reaction without undergoing any change itself

Doping containing a small amount of a material as an impurity

Isotopes two or more forms of an element that differ from each other according to their mass number

Metalloid an element that has characteristics of both metals and a non-metals

Periodic table a chart that shows how the chemical elements are related to each other

Radioactive isotope an isotope that breaks apart and gives off some form of radiation

Semiconductor a material that conducts an electric current, but not nearly as well as metals

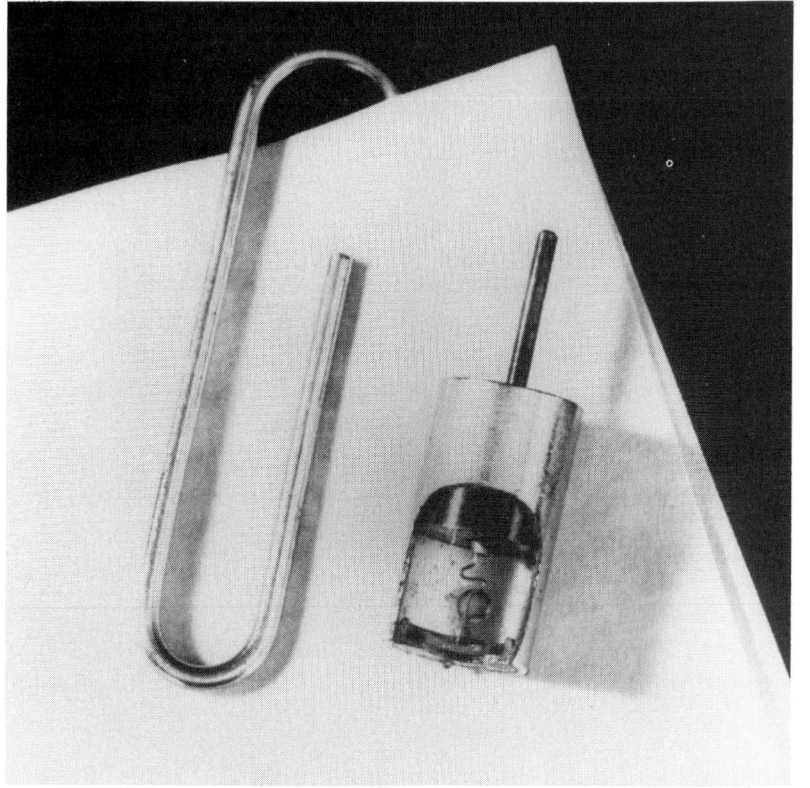

Germanium is used in the production of transistors. This transistor was made in the 1950s.

Physical properties

Germanium looks like a metal. It has a bright, shiny, silvery color. But it is brittle and breaks apart rather easily, which metals normally do not do. It has a melting point of 937.4°C (1,719°F) and a boiling point of 2,830°C (5,130°F). It conducts an electric current poorly. Substances of this kind are called semiconductors. Semiconductors conduct an electric current, but not nearly as well as metals like silver, **copper,** and **aluminum.**

The ability of semiconductors to conduct electricity depends greatly on the presence of small amounts of impurities. The addition of an impurity to a semiconductor is called doping. Doping a semiconductor has significant effects on its ability to conduct an electric current.

Chemical properties

Germanium is a relatively inactive element. It does not dissolve in water and does not react with **oxygen** at room temperature. It does dissolve in hot acids and with oxygen at high temperatures, however. It becomes more active when finely divided. It will combine with **chlorine** and **bromine** to form germanium chloride ($GeCl_4$) and germanium bromide ($GeBr_4$). For example:

$$Ge \text{ (finely divided)} + 2Cl_2 \xrightarrow{\text{high temperature}} GeCl_4$$

Occurrence in nature

The abundance of germanium in the Earth's crust is estimated to be about 7 parts per million. That places it in the bottom third of the elements arranged according to their abundance.

The two most common minerals of germanium are argyrodite and germanite. Argyrodite is the mineral in which Winkler first discovered germanium. Germanite contains about 8 percent germanium.

Most germanium today is obtained from **zinc** ores. When those ores are treated to obtain zinc metal, some germanium is produced at the same time.

It is possible to buy germanium that is 99.9999 percent pure. This level of purity is needed in order to use the metal in the production of semiconductors.

Germanium is obtained from two mines in the United States. One mine is in Alaska and the other is in Tennessee. The United States also imports germanium from China, the United Kingdom, Ukraine, Russia, Belgium, and other nations.

Isotopes

There are five naturally occurring isotopes of germanium: germanium-70, germanium-72, germanium-73, germanium-74, and germanium-76. Isotopes are two or more forms of an element. Isotopes differ from each other according to their mass number. The number written to the right of the element's name is the mass number. The mass number represents the number of protons plus neutrons in the nucleus of an atom of the element. The number of protons determines the element, but the number of neutrons in the atom of any one element can vary. Each variation is an isotope.

At least nine radioactive isotopes of germanium are known also. A radioactive isotope is one that breaks apart and gives off some form of radiation. Radioactive isotopes are produced when very small particles are fired at atoms. These particles stick in the atoms and make them radioactive.

None of the radioactive isotopes of germanium has any important commercial use.

The glass in satellites often contains germanium. This satellite was launched in June 1990.

Extraction

Germanium in zinc ores is heated in the presence of chlorine gas. Germanium chloride ($GeCl_4$) is formed:

$$Ge \text{ (in ores)} + 2Cl_2 \xrightarrow{\text{heat}} GeCl_4$$

Pure germanium metal is then produced by passing an electric current through molten (melted) germanium chloride:

$$GeCl_4 \text{ —electric current→ } Ge + 2Cl_2$$

This method produces very pure germanium. It is possible to buy germanium that is 99.9999 percent pure. This level of purity is needed in order to use the metal in the production of semiconductors.

Uses

Germanium first became important for its use in semiconductors. This still accounts for about 15 percent of the germanium produced. But other uses of the element are now more important. About 40 percent of the germanium produced in the United States is now used in the manufacture of fiber optic systems.

Germanium is used to make weapons-sighting systems that can be used in the dark.

An optical fiber is a very thin thread made out of pure glass. The fiber acts somewhat like a copper wire. It can carry messages in the same way a copper wire carries messages. The difference is that optical fibers carry messages on light waves. Copper wires carry messages on electric currents.

The ability of a glass thread to carry light depends on the presence of impurities. Optical fibers are doped with germanium and other elements to improve their ability to carry light messages. Optical fibers are now used to carry telephone messages instead of electric wires.

Germanium is also used as a catalyst. A catalyst is used to speed up or slow down a chemical reaction. The catalyst does not undergo any chemical change during the reaction. Germanium catalysts are used primarily in the production of plastics.

Germanium is also used to make specialized glass for military applications. For example, it is used to make weapons-sighting systems that can be used in the dark. Satellite systems and fire alarm systems may also contain glass that contains germanium.

Compounds

Few germanium compounds have any important commercial uses.

Health effects

Germanium is not thought to be essential to the health of plants or animals. Some of its compounds present a hazard to

human health, however. For example, germanium chloride and germanium fluoride (GeF_4) are a liquid and gas, respectively, that can be very irritating to the eyes, skin, lungs, and throat.

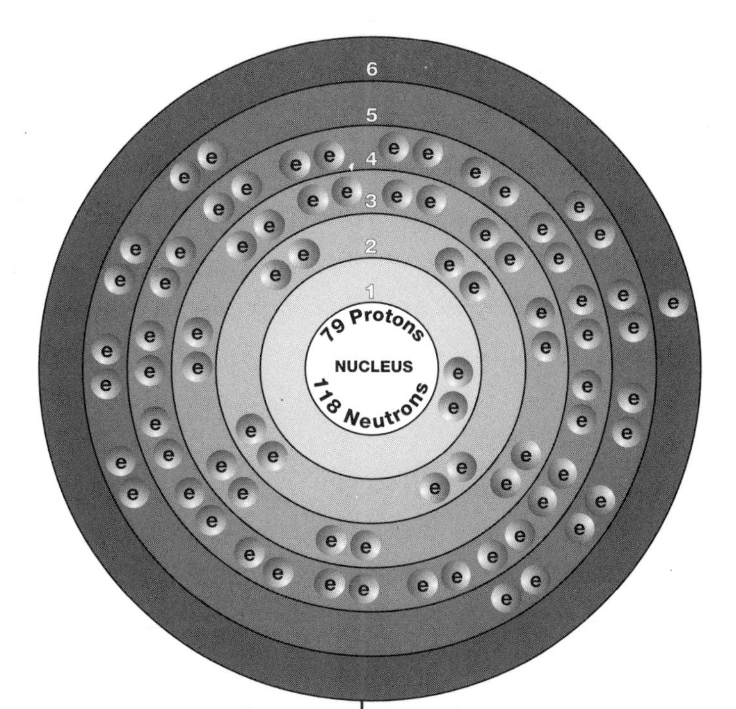

GOLD

Overview

Gold has been called the most beautiful of all chemical elements. Its beauty has made it desirable for use in jewelry, coins, and artwork for thousands of years. It was one of the first pure metals to be used by humans.

Gold is one of the few elements that can affect politics and economics. Wars have been fought over access to gold. Cities and towns have sprung up and died out as gold was discovered and then mined out. Many nations still count their wealth according to the amount of gold they keep in storage.

Gold lies in the middle of the periodic table. The periodic table is a chart that shows how elements are related to one another. Gold is a heavy metal in a group known as the transition metals. Gold is also known as a precious metal (as are **platinum** and **silver**).

Large amounts of gold are still used in the manufacture of coins, medals, jewelry, and art. Gold also has a number of uses in industry, medicine, and other applications. For example, one radioactive isotope of gold is commonly used to treat cancer.

The chemical symbol for gold is Au. The symbol comes from the Latin word for gold, *aurum*. Aurum means "shining dawn."

SYMBOL
Au

ATOMIC NUMBER
79

ATOMIC MASS
196.9665

FAMILY
Group 11 (IB)
Transition metal

PRONUNCIATION
GOLD

Goin' for the silver *and* gold!

One of the most famous items using gold is the Olympic gold medal. Athletes from around the world dream of coming in first place at the Olympics. That means they can step up to the winner's podium and wear their gold medal proudly. But the gold medal isn't solid gold. It's actually made out of silver. A thin layer of gold covers the silver. The last time a solid gold medal was used in the Olympics was 1912.

Christopher Columbus found gold nuggets lying in the bottom of rivers and harbors in Haiti.

Discovery and naming

Gold objects dating to 2600 B.C. have been found. They were discovered in the royal tombs of the ancient civilization of Ur. These objects showed that humans had already learned how to work with gold this early in history. Some of the gold, for example, had been formed into wires.

One of the special skills developed by the Egyptians was the adding of gold to glass objects. They found a way to use gold to make glass a beautiful ruby-red color. The glass became known as gold ruby glass.

Gold is also mentioned in a number of places in the Bible. A passage in Exodus, for example, refers to the clothing worn by Aaron: "And they did beat the gold into thin plates, and cut it into wires, to work it in the blue, and in the purple, and in the scarlet, and in the fine linen, with cunning work."

Writings from every stage of human history tell of the discovery and use of gold. Roman historian Pliny the Elder (A.D. 23–79), for example, describes gold-mining locations. The Romans found it lying in stream beds in the Tagus River in Spain, the Po River in Italy, the Hebrus River in Thracia (now Greece), the Pactolus River in Asia Minor (now Turkey), and the Ganges River in India.

Gold has long been known in the New World, too. During a visit to Haiti, Christopher Columbus (1451–1506) found gold nuggets lying on the bottom of rivers and harbors. A Portuguese explorer in 1586, Lopez Vaz, wrote that the region called Veragua (now Panama) was the "richest Land of Gold [in] all the rest of the Indies."

Olympic champion Tara Lipinski holds up her gold medal.

In the United States, of course, the most famous story about gold occurred in the late 1840s. Thousands of people flocked to California in search of gold. This era was called the Gold Rush. People became very rich or found nothing at all during this exciting time in history.

Physical properties

Gold is both ductile and malleable. Ductile means it can be drawn into thin wires. Malleable means capable of being hammered into thin sheets. A piece of gold weighing only 20 grams (slightly less than an ounce) can be hammered into a sheet that will cover more than 6 square meters (68 square feet). The sheet will be only 0.00025 centimeters (one ten-thousandth of an inch) thick. Gold foil of this thickness is often used to make the lettering on window signs.

Gold is quite soft. It can usually be scratched by a penny. Its melting point is 1,064.76°C (1,948.57°F) and its boiling point is about 2,700°C (4,900°F). Its density is 19.3 grams per cubic centimeter.

Two other important properties are its reflectivity and lack of electrical resistance. Both heat and light reflect off gold very well. But an electric current passes through gold very easily.

Chemical properties

Generally speaking, gold is not very reactive. It does not combine with **oxygen** or dissolve in most acids. It does not react with halogens, such as **chlorine** or **bromine**, very easily.

These chemical properties also account for some important uses of gold. Gold coins, for example, do not corrode (rust) or tarnish very easily. Neither does jewelry or artwork made of gold.

Occurrence in nature

Gold occurs in nature in both its native state and in compounds. The native state of an element is its free state. It is not combined with any other element. The most common compounds of gold are the tellurides. A telluride is a compound of the element **tellurium** and one or more other elements. For example, the mineral calavarite is mostly gold telluride ($AuTe_2$).

At one time, gold was found in chunks or nuggets large enough to see. People mined gold by picking it out of streams and rivers. In fact, gold was once very common in some parts of the world. People valued it not because it was rare, but because it was so beautiful.

The abundance of gold in the Earth's crust is estimated to be about 0.005 parts per million. That makes it one of the ten rarest elements in the Earth's crust. Gold is thought to be much more common in the oceans. Some people believe as

WORDS TO KNOW

Alloy a mixture of two or more metals with properties different from those of the individual metals

Amalgam a mixture of two or more metals, one of which is mercury

Carat a unit of weight for gold and other precious metals, equal to one fifth of a gram, or 200 milligrams

Ductile capable of being drawn into thin wires

Isotopes two or more forms of an element that differ from each other according to their mass number

Malleable capable of being hammered into thin sheets

Radioactive isotope an isotope that breaks apart and gives off some form of radiation

Transition metal an element in Groups 3 through 12 of the periodic table

The Gold Rush!

The most famous story about gold in the United States might be the Gold Rush of 1849. As early as the sixteenth century, records contained stories about a great El Dorado ("the gilded one," in Spanish; gilded means "covered in gold") on the western coast of the United States. Tales of this magical city were repeated for centuries.

In the late 1840s, explorers began to travel from the Eastern seaboard to California in search of El Dorado. The flow of visitors was slow at first. Gold was first discovered in 1848 at a place called Sutter's Mill. Sutter's Mill was located near the present town of Coloma, California.

Word of the discovery spread quickly. Within a year, thousands of men and women made the long, expensive, and tiring trip. Most people traveled across the United States in covered wagons or on horseback. Many of them had to cross mountains, plains, and deserts. Because of the difficult conditions, many people and animals got sick or died. Some people traveled around Cape Horn at the bottom of South America or across the Isthmus of Panama. No matter which route was used, the journey usually took months.

As people arrived in California, hundreds of mining camps sprang up. Some of them had colorful names. Poker Flat, Hangtown, Red Dog, Hell's Delight, and Whiskey Bar were just a few! Mining for gold was hard work. Gold miners usually wound up being wildly successful or terrible failures. The Gold Rush of 1849 completely changed the state of California. It also helped expand the United States.

much as 70 million tons of gold are dissolved in seawater. They also think there may be another 10 billion tons on the bottom of the oceans. So far, however, no one has found a way to mine this gold.

About a quarter of the world's gold comes from South Africa. Other leading producers of the metal are the United States, Australia, Canada, China, and Russia. In the United States, about two-thirds of its gold is mined in Nevada. California, Montana, Alaska, and South Dakota also produce gold.

At one time, gold was found in chunks or nuggets large enough to see.

Isotopes

There is only one naturally occurring isotope of gold, gold-197. Isotopes are two or more forms of an element. Isotopes differ from each other according to their mass number. The number written to the right of the element's name is the mass number. The mass number represents the number of protons plus neutrons in the nucleus of an atom of the element. The number of protons determines the element, but the number of neutrons

Pieces of gold attached to a chunk of quartz.

in the atom of any one element can vary. Each variation is an isotope.

About two dozen radioactive isotopes of gold are known also. A radioactive isotope is one that breaks apart and gives off some form of radiation. Radioactive isotopes are produced when very small particles are fired at atoms. These particles stick in the atoms and make them radioactive.

In a 1986 study, experts estimated that 121,000 tons of gold had been mined throughout history.

One radioactive isotope of gold is widely used in medicine, gold-198. This isotope has two major uses. First, it can be used to study the liver. It is made into a form known as colloidal gold. Colloidal gold consists of very fine particles of gold mixed in a liquid solution. The colloidal gold is injected into the patient's body, where it travels to the liver. There, it can be detected because of the radiation it gives off. The radiation can be used to tell if the liver is functioning normally or not.

Colloidal gold is also used to treat medical problems. In some forms of cancer, the body develops large amounts of liquid in the space around the stomach and intestines (the peritoneum). One way to treat this collection of liquid is with colloidal gold. The colloidal gold is injected into the peritoneum.

Measuring gold

Which weighs more: A pound of feathers or a pound of gold? Teachers sometimes try to fool students with this old question. The answer would seem to be easy: a pound is a pound. A pound of feathers and a pound of gold should weigh the same amount.

But that is not quite true. In the English system, most substances are measured using the avoirdupois (pronounced a-ver-de-POIZ) system. In the avoirdupois system, there are 16 ounces to the pound.

But gold is weighed differently. It uses the troy system. In the troy system, one pound contains only 12 ounces. So, a pound of feathers (avoirdupois system) weighs four ounces more than a pound of gold (troy system). The weight of other precious metals, like silver and platinum, are also measured using the troy system.

Gold is also weighed in carats. A carat is defined as one fifth of a gram, or 200 milligrams.

Gold is seldom used in a pure form. The metal is too soft. It would bend or break if used pure. Instead, it is used in combination with other metals called alloys. An alloy is a mixture of two or more metals. The mixture has properties different from those of the individual metals.

The amount of gold in an alloy is expressed in carats. Pure gold metal (mixed with no other metal) is said to be 24-carat gold. An alloy that contains 20 parts of gold and 4 parts of silver is 20-carat gold. The "20-carat" designation means the alloy contains 20 parts of gold and 4 parts of something else (silver, in this case).

Gold stored in a national bank can be 24-carat gold. It is never used for any practical purpose. But gold used for any real application is almost always less than 24 carats. It must include other metals that make it stronger and tougher.

It is not able to leave the peritoneum and go into the stomach and intestines. While in the peritoneum, the colloidal gold gives off radiation. The radiation kills cancer cells that cause the accumulation of fluid.

Extraction

There are at least two main ways to remove gold from its ores. One is to mix an ore with **mercury** metal. Mercury combines with gold in the ore to form an amalgam. An amalgam is a mixture of two or more metals, one of which is mercury. The gold amalgam is then removed from the ore. It is heated to drive off the mercury. Pure gold remains.

Gold ores can also be treated with potassium cyanide (KCN) or some other kind of cyanide. The gold combines with the cyanide to form a new compound, gold cyanate. The gold cyanate is then treated with an active metal, such as zinc. The active metal replaces gold in the compound, leaving pure gold.

Uses

In a 1986 study, experts estimated that 121,000 tons of gold had been mined throughout history. Of that amount, about 18,000 tons were used for industrial, research, health, and other "dissipative" uses. Dissipative means that the gold was gone once it was used. It was made into devices that were eventually thrown away. The gold could not or was not recovered from the devices.

Of the remaining 103,000 tons of gold, about a third (35,000 tons) had been made into gold bars held by national banks. The gold bars are used as security for national money systems. In the United States, for example, the nation's supply of gold is stored at Fort Knox, Kentucky.

In the United States, the nation's supply of gold is stored at Fort Knox, Kentucky.

Finally, the remaining 68,000 tons of gold are owned by private individuals. This gold exists in the form of jewelry, coins, or bullion. Gold bullion are bars or other large pieces of pure gold.

Jewelry is the largest single use of gold. In 1996, about 3,290 tons of gold were made worldwide. Of that amount, nearly 85 percent was made into jewelry. The second largest use of gold (about 213 tons, or about 7 percent) was in industrial devices and consumer products. Some examples include electrical contacts and switches, laboratory equipment, printed circuits, dental alloys, instruments on space vehicles, and nozzles used in the production of synthetic fibers.

Compounds

Few gold compounds have any important commercial uses.

Health effects

Gold is not required to maintain good health in plants or animals. It can be injected into a plant or animal without causing harmful effects. Some medical and commercial uses are based on this property.

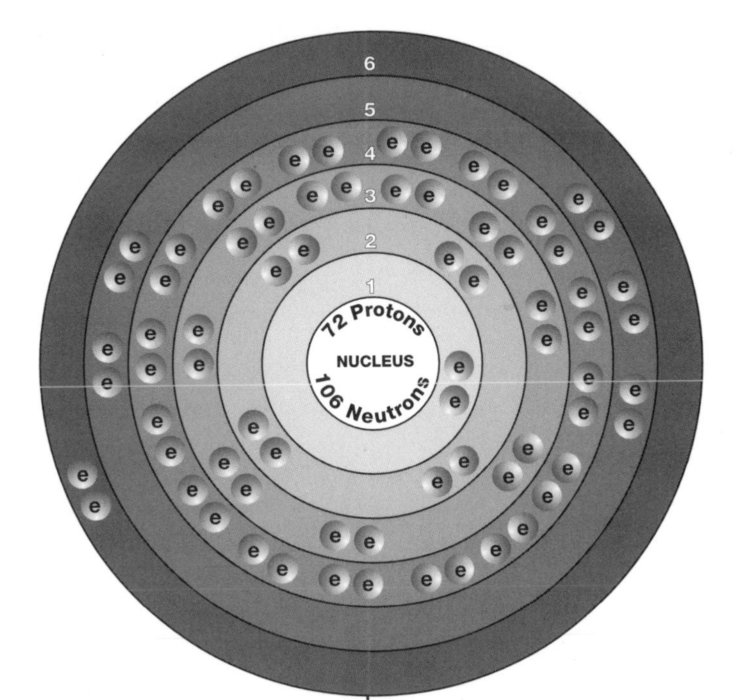

HAFNIUM

Overview

Hafnium is an element that chemists knew existed, but could not find. They knew it must exist because of an empty space in the periodic table. The periodic table is a chart that shows how chemical elements are related to each other.

By the twentieth century, nearly all the spaces in the periodic table had been filled. One of the empty spaces was element 72. A number of chemists searched for the element. Some even claimed they had found it. But these claims were not confirmed. In fact, it was not until 1923 that element number 72 was finally discovered.

Hafnium is a shiny, silvery-white metal. It is always found with another chemical element, **zirconium.** The two elements are very much alike. In fact, their similarity is the reason that it took so long to find hafnium.

Hafnium has only a few applications. Probably its most important use is in nuclear power plants. A nuclear power plant is a facility where energy released from nuclear fission reactions is used to generate electricity.

SYMBOL
Hf

ATOMIC NUMBER
72

ATOMIC MASS
178.49

FAMILY
Group 4 (IVB)
Transition metal

PRONUNCIATION
HAF-nee-um

Discovery and naming

In the early 1900s, scientists had found a new way to identify elements. This method is called X-ray diffraction analysis. Here is how this method works:

A stream of electrons is fired at a metal plate. The electrons can cause the metal plate to give off X rays. This is a simple explanation of what happens when a person has a chest X ray or an X ray of a broken bone.

The kind of X ray produced depends on the metal used. Each metal produces its own special X-ray pattern. In fact, the pattern produced can be used to identify a metal.

In 1923 Dutch physicist Dirk Coster (1889–1950) and Hungarian chemist George Charles de Hevesy (1889–1966) found element 72 by X-ray analysis. The element was present in a piece of Norwegian zircon. Zircon also contains the mineral zirconium.

Chemists later developed a better understanding about the relationship of zirconium and hafnium. These two elements are as alike as any two elements in the periodic table. They have nearly identical chemical and physical properties. This explains why it took so long to find hafnium. Chemists had probably discovered hafnium before 1923, but thought it was zirconium. The differences in the X-ray patterns of the two elements finally proved that hafnium was different from zirconium.

Physical properties

Hafnium is a bright, silvery-gray metal that is very ductile. Ductile means capable of being drawn into thin wires. Its melting point is about 2,150°C (3,900°F) and its boiling point about 5,400°C (9,700°F). Its density is 13.1 grams per cubic centimeter.

The physical property of greatest interest for hafnium is how it responds to neutrons. A neutron is a very small particle found in the nucleus (center) of an atom. Neutrons are used to make nuclear fission reactions occur. Nuclear fission reactions take place when a neutron strikes a large atom, such as an atom of **uranium.** The neutron makes the atom break apart. In the process, a large amount of energy is released. That energy can be converted to electricity.

In order to make electricity from nuclear fission, the fission reaction must be carefully controlled. To do that, the number of

WORDS TO KNOW

Alloy a mixture of two or more metals with properties different from those of the individual metals

Ductile capable of being drawn into thin wires

Half life the time it takes for half of a sample of a radioactive element to break down

Isotopes two or more forms of an element that differ from each other according to their mass number

Periodic table a chart that shows how the chemical elements are related to each other

Radioactive isotope an isotope that breaks apart and gives off some form of radiation

Hafnium samples.

neutrons must also be kept under close control. Hafnium has the ability to absorb ("soak up") neutrons very easily. It is used in rods that control how fast a fission reaction takes place.

This property is one of the few ways in which hafnium differs from zirconium. While hafnium is very good at absorbing neutrons, zirconium hardly absorbs neutrons at all. Neutrons pass right through it. Both hafnium and zirconium are used in nuclear power plants.

Hafnium and zirconium are as alike as any two elements in the periodic table.

Chemical properties

Like zirconium, hafnium is not very reactive. It does not combine easily with **oxygen** in the air or react with water or cold acids. It may be more active with hot acids, however.

Occurrence in nature

Hafnium is a moderately common element in the Earth's crust. Its abundance is estimated to be about 5 parts per million. That makes it about as abundant as **bromine,** uranium, or **tin.**

Hafnium is used in nuclear power plants. Pictured here is the Davis-Besse Nuclear Power Plant in Oak Harbor, Ohio.

Hafnium is always found with ores of zirconium in the earth. The most common of those ores are zircon and baddeleyite.

Isotopes

Hafnium has six naturally occurring isotopes: hafnium-174, hafnium-176, hafnium-177, hafnium-178, hafnium-179, and hafnium-180. The first of these is radioactive. It has a half life of an astounding two quadrillion years. (That's 2 followed by 15 zeroes!) Isotopes are two or more forms of an element. Isotopes differ from each other according to their mass number. The number written to the right of the element's name is the mass number. The mass number represents the number of protons plus neutrons in the nucleus of an atom of the element. The number of protons determines the element, but the number of neutrons in the atom of any one element can vary. Each variation is an isotope.

Artificial radioactive isotopes can also be produced by firing very small particles at atoms. These particles stick in the atoms and make them radioactive. About 12 artificial radioactive isotopes of hafnium are known. There are no important commercial applications for any of these isotopes.

Extraction

The greatest problem in working with hafnium is finding a way to separate it from zirconium. Today, chemists know that compounds of hafnium dissolve more easily in some liquids than do compounds of zirconium. This method can be used to separate compounds of the two elements from each other.

Uses and compounds

Nuclear power plant applications account for the largest use of hafnium metal. Hafnium is also used to make binary compounds with interesting properties. A binary compound consists of two elements. These compounds are among the best refractory materials known. A refractory material is one that can withstand very high temperatures. It reflects heat away from itself. Refractory materials are used to line the inside of high-temperature ovens. For example, some alloys are made at temperatures of thousands of degrees in refractory ovens. Some hafnium compounds used to line these furnaces are hafnium boride (HfB_2), hafnium nitride (HfN), and hafnium oxide (HfO_2).

Nuclear power plant applications account for the largest use of hafnium metal.

Health effects

Both hafnium and its compounds are toxic. They are most dangerous when inhaled. Powdered hafnium metal is also dangerous. It can ignite and explode very easily.

HELIUM

Overview

Helium is a member of the noble gas family. The noble gases are the elements in Group 18 (VIIIA) of the periodic table. The periodic table is a chart that shows how the elements are related to one another. The noble gases are also called the inert gases. Inert means that an element is not very active. It will not combine with other elements or compounds. In fact, no compounds of helium have ever been made.

Helium is the second most abundant element in the universe. Only **hydrogen** occurs more often than helium. Helium is also the second simplest of the chemical elements. Its atoms consist of two protons, two neutrons, and two electrons. Only the hydrogen atom is simpler than a helium atom. The hydrogen atom has one proton, one electron, and no neutrons.

Helium was first discovered not on Earth, but in the Sun. In 1868 French astronomer Pierre Janssen (1824–1907) studied light from the Sun during a solar eclipse. He found proof that a new element existed in the Sun. He called the element helium.

Helium has some interesting and unusual physical properties. For example, at very low temperatures it can become superflu-

SYMBOL
He

ATOMIC NUMBER
2

ATOMIC MASS
4.002602

FAMILY
Group 18 (VIIIA)
Noble gas

PRONUNCIATION
HEE-lee-um

Weather balloons are often filled with helium.

id. A superfluid material behaves very strangely. It can flow upwards out of a container, against the force of gravity. It can also squeeze through very small holes that should be able to keep it out. The Nobel Prize in physics for 1996 was awarded to three Americans who discovered superfluidity. They were David M. Lee (1931-), Douglas D. Osheroff (1945-), and Robert C. Richardson (1937-).

For an inactive gas, helium has a surprising number of applications. It is used in low-temperature research, for filling balloons and dirigibles (blimps), to pressurize rocket fuels, in welding operations, in lead detection systems, in **neon** signs, and to protect objects from reacting with **oxygen.**

Discovery and naming

One of the most powerful instruments for studying chemical elements is the spectroscope. The spectroscope is a device for studying the light produced by a heated object. For example, a lump of **sodium** metal will burn with a yellow flame. The flame looks quite different, however, when viewed through a spectroscope.

A spectroscope contains a triangular piece of glass (called a prism) that breaks light into its basic parts. These basic parts consist of a series of colored lines. In the case of sodium, the yellow light is broken into a series of yellow lines. These lines are called the element's *spectrum.* Every element has its own distinctive spectrum.

French astronomer Pierre Janssen discovered helium after studying the Sun during a full solar eclipse.

The spectroscope gives scientists a new way of studying elements. They can identify an element by recognizing its distinctive spectral lines even when they can't actually see the element itself.

This principle led to the discovery of helium. In 1868, Janssen visited India in order to observe a full eclipse of the Sun. A solar eclipse occurs when the Moon comes between the Sun and the Earth. The Moon blocks nearly all of the Sun's light. All that remains is a thin outer circle (corona) of sunlight around the dark Moon. Solar eclipses provide scientists with an unusual chance to study the Sun.

Janssen examined light from the Sun with a spectroscope. As he looked at the spectral lines, he was surprised to see some lines that could not be traced to any known element. He concluded that there must be an element on the Sun that had never been seen on Earth. The name helium was later suggested for this element. The name comes from the Greek word *helios* for "sun."

Chemists did not know what to make of Janssen's discovery. Was there an element on the Sun that did not exist on Earth? Had he made a mistake? Some scientists even made fun of Janssen.

Ernest Rutherford | English physicist

"That's the last potato I'll ever dig!" That statement was attributed to young Ernest Rutherford, a native of New Zealand. Rutherford had applied for a scholarship to Cambridge University, one of England's most famous universities. Rutherford had actually finished second in the scholarship competition. But the winner had decided to stay in New Zealand and get married. When Rutherford was told he had won the scholarship, he threw down his potato fork and announced the end of his potato-digging days.

Rutherford went on to become one of the great scientific figures of the twentieth century. He made a number of important discoveries about the structure of atoms and about radioactivity. For example, he found that an atom consists of two distinct parts, the nucleus and the electrons. He also discovered one form of radiation given off by radioactive materials: alpha particles. Alpha particles, he found, are simply helium atoms without their electrons.

Rutherford was also the first scientist to change one element into another. He accomplished this by bombarding nitrogen gas with alpha particles. Rutherford found that oxygen was formed in this experiment. He had discovered a way to convert one element (nitrogen) into a different element (oxygen). The method Rutherford used later became a standard procedure used by many other scientists.

For the next thirty years, chemists looked for helium on Earth. Then, in 1895, the English physicist Sir William Ramsay (1852–1916) found helium in a mineral of the element **uranium.** Credit for the earthly discovery of helium is sometimes given to two other scientists also. Swedish chemists Per Teodor Cleve (1840–1905) and Nils Abraham Langlet also discovered helium at about the same time in a mineral called cleveite.

Ramsay did not know why helium occurred in an ore of uranium. Some years later, the reason for that connection became obvious. Uranium is a radioactive element. A radioactive element is one that breaks apart spontaneously. It releases radiation and changes into a new element.

One form of radiation produced by uranium consists of alpha particles. Alpha particles are tiny particles moving at very high

rates of speed. In 1907, English physicist Ernest Rutherford (1871–1937) showed that an alpha particle is nothing more than a helium atom without its electrons. As uranium atoms broke apart, then, they gave off alpha particles (helium atoms). That is why helium was first found on Earth in connection with uranium ores.

Physical properties

Helium is a colorless, odorless, tasteless gas. It has a number of unusual properties. For example, it has the lowest boiling point of any element, -268.9°C (-452.0°F). The boiling point for a gas is the temperature at which the gas changes to a liquid. The freezing point of helium is -272.2°C (-458.0°F). Helium is the only gas that cannot be made into a solid simply by lowering the temperature. It is also necessary to increase the pressure on the gas in order to make it a solid.

At a temperature of about -271°C (-456°F), helium undergoes an unusual change. It remains a liquid, but a liquid with strange properties. Superfluidity is one of these properties. The forms of helium are so different that they are given different names. Above -271°C, liquid helium is called helium I; below that temperature, it is called helium II.

Chemical properties

Helium is completely inert. It does not form compounds or react with any other element.

Occurrence in nature

Helium is the second most abundant element after hydrogen in the universe and in the solar system. About 11.3 percent of all atoms in the universe are helium atoms. By comparison, about 88.6 percent of all atoms in the universe are hydrogen. Thus, at least 99.9 percent of all atoms are either hydrogen or helium atoms.

By contrast, helium is much less abundant on Earth. It is the sixth most abundant gas in the atmosphere after **nitrogen,** oxygen, **argon,** carbon dioxide, and neon. It makes up about 0.000524 percent of the air.

It is probably impossible to estimate the amount of helium in the Earth's crust. The gas is produced when uranium and other

WORDS TO KNOW

Alpha particles helium atoms without their electrons

Inert not very active

Isotopes two or more forms of an element that differ from each other according to their mass number

Liquid air air that has been cooled to a very low temperature

Noble gas an element in Group 18 (VIIIA) of the periodic table

Periodic table a chart that shows how the chemical elements are related to each other

Radioactive isotope an isotope that breaks apart and gives off some form of radiation

Spectroscope a device for studying the light produced by a heated object

Superconductor a material that has no resistance to the flow of electricity; once an electrical current begins flowing in the material, it continues to flow forever

radioactive elements break down. But it often escapes into the atmosphere almost immediately.

Isotopes

Two isotopes of helium occur naturally, helium-3 and helium-4. Isotopes are two or more forms of an element. Isotopes differ from each other according to their mass number. The number written to the right of the element's name is the mass number. The mass number represents the number of protons plus neutrons in the nucleus of an atom of the element. The number of protons determines the element, but the number of neutrons in the atom of any one element can vary. Each variation is an isotope.

Helium often occurs along with natural gas in reservoirs deep beneath the Earth's surface.

Three radioactive isotopes of helium have been made also. A radioactive isotope is one that breaks apart and gives off some form of radiation. Radioactive isotopes are produced when very small particles are fired at atoms. These particles stick in the atoms and make them radioactive.

None of the radioactive isotopes of helium has any commercial application.

Extraction

In theory, helium could be collected from liquid air. Liquid air is air that has been cooled to a very low temperature. All of the gases in air have liquefied in liquid air. If the liquid air were allowed to evaporate, the last gas remaining after all other gases had evaporated would be helium. There is too little helium in air to make this process worthwhile, however.

There is a much better source of helium. The gas often occurs along with natural gas in reservoirs deep beneath the Earth's surface. When wells are dug to collect the natural gas, helium comes to the surface with the natural gas. Then, helium can be separated from natural gas very easily. The temperature of the mixture is lowered, and the natural gas liquefies and is taken away. Gaseous helium is left behind.

About 80 percent of the world's helium is in the United States. In 1996, 20 U.S. plants produced helium from gas wells. About 86 percent of U.S. helium comes from five large underground regions: the Hugoton Field that lies beneath Kansas, Oklahoma, and Texas; the Keyes Field in Oklahoma; the Panhandle and Cliffside Fields in Texas; and the Ridley Ridge area in Wyoming.

For many years, the U.S. government operated the Federal Helium Program. This program was responsible for collecting and storing helium for government use. The main customers for this helium were the Department of Defense, the National Aeronautics and Space Administration, and the Department of Energy. The helium was stored underground in huge natural caves.

In 1996, the government decided to end this program. Helium was no longer regarded as essential to national security. The Bureau of Mines has begun to sell off the federal reserves.

Uses

The most important single use for helium in the United States is in low-temperature cooling systems. This is because liquid helium—at -270°C—is cold enough to cool *anything* else. For example, it is used in superconducting devices.

Helium is used to inflate balloons and other lighter-than-air crafts, such as dirigibles (blimps).

A superconducting material is one that has no resistance to the flow of electricity. Once an electric current begins to flow in the material, it will continue to flow forever. No energy is wasted in moving electricity from one place to another. Superconducting materials may revolutionize electrical systems worldwide someday. The problem is that superconductivity occurs only at very low temperatures. One way to achieve those temperatures is with liquid helium.

Another important use of helium is in pressure and purge systems. In many industrial operations, it is necessary to pressurize a system. The easiest way to do that is to pump a gas into the system. But the gas should not be one that will react with other substances in the system. Being inert, helium is a perfect choice. Helium is also used for purging, a process that sweeps away all gas in a container. Again, helium is used because it does not react with anything in the container.

Because of its inactivity, helium is also used in welding systems. Welding is the process by which two metals are heated to high temperatures in order to join them to each other. Welding rarely works well in "normal" air. At high temperatures, the metals may react with oxygen to form metal oxides. If they do, they are less likely to join to each other. If the welding is done in a container of helium, this is not a problem. The metals will not react with helium. They will simply join to each other.

Helium is used to inflate dirigibles, such as the Goodyear blimps.

Helium is also used in leak-detection systems. If a leak is suspected in a long pipe, helium can be used to look for that leak. It is pumped into one end of the pipe. A detector is held outside the pipe. The detector is designed to measure whether helium is escaping from the system. The detector is moved along the length of the pipe. It is possible to find out whether there is a leak, where it is, and how much it is leaking. Helium is a good gas to use for this purpose because it does not react with anything in the pipe.

Helium is also used to inflate balloons and other lighter-than-air crafts, such as dirigibles (blimps). Helium does not have the lifting power of hydrogen. However, hydrogen is flammable and helium is not. At one time, people thought that dirigibles would be a popular form of transportation. But that never happened. Blimps are still used for limited purposes, such as advertising at major sports and recreational events.

Compounds

No compounds of helium have ever been made.

Health effects

There are no known health hazards resulting from exposure to helium.

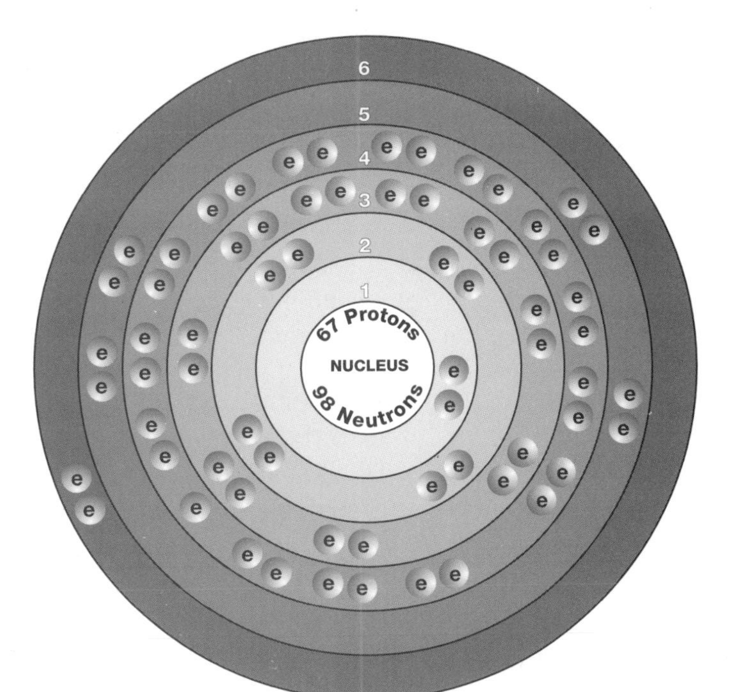

HOLMIUM

Overview

Holmium occurs in Row 6 of the periodic table. The periodic table is a chart that shows how chemical elements are related to each other. Elements with atomic numbers 58 through 71 are known as the lanthanides. The name comes from the element **lanthanum.** The lanthanides are also known as rare earth elements. Although lanthanides are not especially rare, they are very difficult to separate from each other.

Holmium was discovered by Swedish chemist Per Teodor Cleve (1840–1905) in 1879. He named the element after his birthplace, Stockholm, Sweden. Holmium occurs with other rare earth elements in minerals such as monazite and gadolinite. It can now be separated from other rare earth elements somewhat easily. But no major uses have been found for it or its compounds.

Discovery and naming

In 1787, a lieutenant in the Swedish army named Carl Axel Arrhenius (1757–1824) was exploring a mine near Ytterby, Sweden. Arrhenius was a "rock hound," a person interested in the study of rocks and minerals. In his explorations, Arrhenius found a rock he had never seen before. He asked his friend

SYMBOL
Ho

ATOMIC NUMBER
67

ATOMIC MASS
164.9303

FAMILY
Lanthanide
(rare earth metal)

PRONUNCIATION
HOL-me-um

Johan Gadolin (1760–1852), professor of chemistry at the University of Åbo in Finland, to study it. Gadolin discovered in the rock a new mineral, which was given the name ytterite.

Ytterite proved to be a fascinating puzzle for chemists. The mineral contained a number of different "earths." In chemistry, the term earth refers to a naturally occurring compound of an element. For example, magnesia is a naturally occurring compound—an earth—of the element **magnesium.**

Chemists found the earths in ytterite all had very similar properties. However, they had trouble separating them from each other. In fact, it took more than a century to analyze ytterite completely.

In 1879, Cleve was studying an earth taken from yttria called erbia. Erbia had been regarded as a new element for some time. But Cleve separated erbia into three different parts. He called them erbia, holmia, and thulia. Holmia is the earth from which the element holmium comes. For his work, Cleve is given credit for the discovery of holmium.

In Cleve's time, chemical equipment was not very advanced. Chemists usually could not prepare very pure samples of materials. Ten years after the "discovery" of holmium, chemists realized it was actually holmium mixed with another new element, **dysprosium.**

Physical properties

Like other rare earth elements, holmium is a silvery metal that is soft, ductile, and malleable. Ductile means capable of being drawn into thin wires. Malleable means capable of being hammered into thin sheets. Both properties are common for metals. Holmium also has some rather unusual magnetic and electrical properties.

Holmium has a melting point of 1,470°C (2,680°F) and a boiling point of 2,720°C (4,930°F). Its density is 8.803 grams per cubic centimeter.

Chemical properties

Holmium metal tends to be stable at room temperature. In moist air and at higher temperatures, it becomes more reactive. For example, it combines with **oxygen** to form holmium

WORDS TO KNOW

Ductile capable of being drawn into thin wires

Earth in mineralogy, a naturally occurring form of an element, often an oxide of the element

Isotopes two or more forms of an element that differ from each other according to their mass number

Lanthanides the elements in the periodic table with atomic numbers between 58 and 71

Laser a device for producing very bright light of a single color

Malleable capable of being hammered into thin sheets

Periodic table a chart that shows how the chemical elements are related to each other

Rare earth elements *see* **Lanthanides**

oxide (Ho_2O_3), a yellow solid. Like most other metals, the element also dissolves in acids.

Occurrence in nature

The abundance of holmium in the Earth's crust is estimated to be about 0.7 to 1.2 parts per million. It is less common than most other rare earth elements, but more common than **iodine, silver, mercury,** and **gold.** The most common ores of holmium are monazite and gadolinite.

Isotopes

Only one naturally occurring isotope of holmium exists, holmium-165. Isotopes are two or more forms of an element. Isotopes differ from each other according to their mass number. The number written to the right of the element's name is the mass number. The mass number represents the number of protons plus neutrons in the nucleus of an atom of the element. The number of protons determines the element, but the number of neutrons in the atom of any one element can vary. Each variation is an isotope.

At least 20 radioactive isotopes of holmium are known also. A radioactive isotope is one that breaks apart and gives off some form of radiation. Radioactive isotopes are produced when very small particles are fired at atoms. These particles stick in the atoms and make them radioactive.

Nonoe of the radioactive isotopes of holmium has any practical uses.

Extraction

Pure holmium is obtained by reacting calcium metal with holmium fluoride (HoF_3):

$$3Ca + 2HoF_3 \rightarrow 3CaF_2 + 2Ho$$

Pure holmium has been available only very recently.

Uses

In the past there were almost no practical uses for holmium and its compounds. However, holmium is now used in specialized lasers. A laser is a device for producing very bright light of a single color. The kind of light produced in a laser depends on the elements of which it is made. Holmium lasers are used to

Holmium lasers are used to reduce abnormal eye pressure and to treat glaucoma.

reduce abnormal eye pressure, to treat glaucoma (an eye disorder), and to repair failed glaucoma surgeries.

Compounds

Few holmium compounds have any important commercial uses.

Health effects

Almost nothing is known about the health effects of holmium. In such cases, the usual recommendation is to treat the element as if it were highly toxic.

Overview

Hydrogen is the most abundant element in the universe. Nearly nine out of every ten atoms in the universe are hydrogen atoms. Hydrogen is also common on the Earth. It is the third most abundant element after **oxygen** and **silicon.** About 15 percent of all the atoms found on the Earth are hydrogen atoms.

Hydrogen is also the simplest of all elements. Its atoms consist (usually) of one proton and one electron.

Hydrogen was first discovered in 1766 by English chemist and physicist Henry Cavendish (1731–1810). Cavendish was also the first person to prove that water is a compound of hydrogen and oxygen.

Some experts believe that hydrogen forms more compounds than any other element. These compounds include water, sucrose (table sugar), alcohols, vinegar (acetic acid), household lye (sodium hydroxide), drugs, fibers, dyes, plastics, and fuels.

SYMBOL
H

ATOMIC NUMBER
1

ATOMIC MASS
1.00794

FAMILY
Group 1 (IA)

PRONUNCIATION
HY-dru-jin

Discovery and naming

Hydrogen was probably "discovered" many times. Many early chemists reported finding a "flammable gas" in some of their experiments. In 1671, for example, English chemist Robert Boyle (1627–91) described experiments in which he added **iron** to hydrochloric acid (HCl) and sulfuric acid (H_2SO_4). In both cases, a gas that burned easily with a pale blue flame was produced.

The problem with these early discoveries was that chemists did not understand the nature of gases very well. They had not learned that there are many kinds of gases. They thought that all the "gases" they saw were some form of air with impurities in it.

Cavendish discovered hydrogen in experiments like those that Boyle performed. He added iron metal to different acids and found that a flammable gas was produced. But Cavendish thought the flammable gas came from the iron and not from the acid. Chemists later showed that iron is an element and does not contain hydrogen or anything else. Therefore, the hydrogen in Cavendish's experiment came from the acid:

$$Fe + 2HCl \rightarrow FeCl_2 + H_2$$

Hydrogen was named by French chemist Antoine-Laurent Lavoisier (1743–94). Lavoisier is sometimes called the father of modern chemistry because of his many contributions to the science. Lavoisier suggested the name hydrogen after the Greek word for "water former" (that which forms water). (See sidebar on Lavoisier in the **oxygen** entry in volume 2.)

WORDS TO KNOW

Catalyst a substance used to speed up or slow down a chemical reaction without undergoing any change

Isotopes two or more forms of an element that differ from each other according to their mass number

Periodic table a chart that shows how the chemical elements are related to each other

Radioactive isotope an isotope that breaks apart and gives off some form of radiation

Tracer a radioactive isotope whose presence in a system can easily be detected

Physical properties

Hydrogen is a colorless, odorless, tasteless gas. Its density is the lowest of any chemical element, 0.08999 grams per liter. By comparison, a liter of air weighs 1.29 grams, 14 times as much as a liter of hydrogen.

Hydrogen changes from a gas to a liquid at a temperature of −252.77°C (−422.99°F) and from a liquid to a solid at a temperature of −259.2°C (−434.6°F). It is slightly soluble in water, alcohol, and a few other common liquids.

Chemical properties

Hydrogen burns in air or oxygen to produce water:

$$2H_2 + O_2 \rightarrow 2H_2O$$

Stars use hydrogen as a fuel with which to produce energy. Antares—the brightest star in the constellation Scorpius—is shown here.

It also combines readily with other non-metals, such as **sulfur, phosphorus,** and the halogens. The halogens are the elements that make up Group 17 (VIIA) of the periodic table. They include **fluorine, chlorine, bromine, iodine,** and **astatine.** As an example:

$$H_2 + Cl_2 \rightarrow 2HCl$$

Occurrence in nature

Hydrogen occurs throughout the universe in two forms. First, it occurs in stars. Stars use hydrogen as a fuel with which to produce energy. The process by which stars use hydrogen is known as fusion. Fusion is the process by which two or more small atoms are pushed together to make one large atom. In most stars, the primary fusion reaction that occurs is:

$$4H \rightarrow 1He$$

This equation shows that four hydrogen atoms are squeezed together (fused) to make one helium atom. In this process,

enormous amounts of energy are released in the form of heat and light.

Hydrogen also occurs in the "empty" spaces between stars. At one time, scientists thought that this space was really empty, that it contained no atoms of any kind. But, in fact, this interstellar space (space between stars) contains a small number of atoms, most of which are hydrogen atoms. A cubic mile of interstellar space usually contains no more than a handful of hydrogen and other atoms.

Hydrogen occurs on the Earth primarily in the form of water. Every molecule of water (H_2O) contains two hydrogen atoms and one oxygen atom. Hydrogen is also found in many rocks and minerals. Its abundance is estimated to be about 1,500 parts per million. That makes hydrogen the tenth most abundant element in the Earth's crust.

The man who gave hydrogen its name, Antoine-Laurent Lavoisier, is sometimes called the father of modern chemistry.

Hydrogen also occurs to a very small extent in the Earth's atmosphere. Its abundance there is estimated to be about 0.000055 percent. Hydrogen is not abundant in the atmosphere because it has such a low density. The Earth's gravity is not able to hold on to hydrogen atoms very well. They float away into outer space very easily. Most of the hydrogen that was once in the atmosphere has now escaped into outer space.

Isotopes

There are three isotopes of hydrogen, hydrogen-1, hydrogen-2, and hydrogen-3. Isotopes are two or more forms of an element. Isotopes differ from each other according to their mass number. The number written to the right of the element's name is the mass number. The mass number represents the number of protons plus neutrons in the nucleus of an atom of the element. The number of protons determines the element, but the number of neutrons in the atom of any one element can vary. Each variation is an isotope.

The three isotopes of hydrogen have special names. Hydrogen-1 is sometimes called protium. It is the simplest and most common form of hydrogen. Protium atoms all contain one proton and one electron. About 99.9844 percent of the hydrogen in nature is protium.

Hydrogen-2 is known as deuterium. A deuterium atom contains one proton, one electron, and one neutron. About 0.0156 percent of the hydrogen in nature is deuterium.

The third isotope of hydrogen, hydrogen-3, is tritium. An atom of tritium contains one proton, one electron, and two neutrons. There are only very small traces of tritium in nature.

Tritium is a radioactive isotope. A radioactive isotope is one that breaks apart and gives off some form of radiation. Some radioactive isotopes (such as tritium) occur in nature. They can also be produced in the laboratory. Very small particles are fired at atoms. These particles stick in the atoms and make them radioactive. Tritium is a widely used isotope and is now made in large amounts in the laboratory.

Tritium is widely used as a tracer in both industry and research. A tracer is a radioactive isotope whose presence in a system can easily be detected. The isotope is injected into the system at some point. Inside the system, the isotope gives off radiation. That radiation can be followed by means of detectors placed around the system.

Stars use hydrogen as a fuel with which to produce energy.

Tritium is popular as a tracer because hydrogen occurs in so many different compounds. For example, suppose a scientist wants to trace the movement of water through soil. The scientist can make up a sample of water made with tritium instead of protium. As that water moves through the soil, its path can be followed by means of the radioactivity the tritium gives off.

Tritium is also used in the manufacture of fusion bombs. A fusion bomb is also known as a hydrogen bomb. In a fusion bomb, small atoms are squeezed together (fused) to make a larger atom. In the process, enormous amounts of energy are given off. For example, the first fusion bomb tested by the United States in 1952 had the explosive power of 15 million tons of TNT. A type of fusion bomb fuses tritium with deuterium to make helium atoms:

$$^{3}H + ^{2}H \rightarrow He$$

Extraction

The obvious source for hydrogen is water. The Earth has enough water to supply people's need for hydrogen. The problem is that it takes a lot of energy to split a water molecule:

$$2H_2O \xrightarrow{\text{electrical current}} 2H_2 + O_2$$

In fact, it simply costs too much to make hydrogen by this method. The cost of electricity is too high. So it is not economical to make hydrogen by splitting water.

A number of other methods can be used to produce hydrogen, however. For example, steam can be passed over hot charcoal (nearly pure **carbon**):

$$H_2O + C \xrightarrow{heat} CO + H_2$$

The same reaction can be used with steam and other carbon compounds. For example, using methane, or natural gas (CH_4), the reaction is:

$$H_2O + CH_4 \xrightarrow{heat} CO + 3H_2$$

Tritium (hydrogen-3, the third isotope of hydrogen), is used in the manufacture of fusion bombs.

Hydrogen can also be made by the reaction between carbon monoxide (CO) and steam:

$$CO + H_2O \xrightarrow{heat} CO_2 + H_2$$

Because hydrogen is such an important element, many other methods for producing it have been invented. However, the preceding methods are the least expensive.

Uses

The most important single use of hydrogen is in the manufacture of ammonia (NH_3). Ammonia is made by combining hydrogen and nitrogen at high pressure and temperature in the presence of a catalyst. A catalyst is a substance used to speed up or slow down a chemical reaction. The catalyst does not undergo any change during the reaction:

$$N_2 + 3H_2 \xrightarrow{pressure,\ heat,\ catalyst} 2NH_3$$

Ammonia is a very important compound. It is used in making many products, the most important of which is fertilizer.

Hydrogen is also used for a number of similar reactions. For example, it can be combined with carbon monoxide to make methanol—methyl alcohol, or wood alcohol (CH_3OH):

$$CO + H_2 \xrightarrow{with\ catalyst} CH_3OH$$

Like ammonia, methanol has a great many practical uses in a variety of industries. The most important use of methanol is in the manufacture of other chemicals, such as those from which

The *Hindenburg* explosion

The *Hindenburg* was Germany's largest passenger airship. It was built in 1936 as a luxury liner, and made the trip to the United States faster than an ocean liner.

The *Hindenburg* was designed to be filled with helium, a safer gas than the highly flammable hydrogen. But in those post-World War II days, the United States suspected that Germany's new leader, Adolf Hitler (1889–1945), had military plans for helium-filled ships. So the United States refused to sell helium to the Zeppelin airship company. Seven million cubic feet of hydrogen was used instead. This made the crew very nervous about the potential for fire. Passengers were even checked for matches as they boarded!

On May 3, 1937, the *Hindenburg* left Frankfurt, Germany, for Lakehurst, New Jersey. It travelled over the Netherlands, down the English Channel, through Canada, and into the United States. Bad weather forced the ship to slow down several times, lengthening the trip. But it finally approached the field in Lakehurst around 7:00 P.M. on May 6. After several minutes of maneuvers due to rain and wind, crewmen dropped ropes to the ground at 7:21. The ship was 200 feet above ground. Four minutes later, a small flame emerged on the skin of the ship, and crewmen heard a pop and felt a shudder. Seconds later, the *Hindenburg* exploded. Flaming hydrogen blasted out of the top. Within 32 seconds, the entire airship had burned, the framework had collapsed, and the entire ship lay smoldering on the ground. Thirty-six people died. Amazingly, 62 survived.

Although claims of sabotage have always surrounded the *Hindenburg* tragedy, American and German investigators both agreed it was an accident. Both sides concluded that the airship's hydrogen was ignited probably by some type of atmospheric electric discharge. Witnesses had noticed some of the skin of the ship flapping; they also observed the nose of the ship rise suddenly. Both indicate the likelihood that free hydrogen had escaped. The *Hindenburg* disaster ended lighter-than-air airship travel for many decades.

plastics are made. Small amounts are used as additives to gasoline to reduce the amount of pollution released to the environment. Methanol is also used widely as a solvent (to dissolve other materials) in industry.

Another important use of hydrogen is in the production of pure metals. Hydrogen gas is passed over a hot metal oxide to produce the pure metal. For example, **molybdenum** can be prepared by passing hydrogen over hot molybdenum oxide:

$$MoO_2 + 2H_2 \xrightarrow{heat} 2H_2O + Mo$$

Hydrogenation is an important procedure to the food industry. In hydrogenation, hydrogen is chemically added to another

The dramatic explosion of the* Hindenburg *in 1937 occurred when hydrogen was ignited.

substance. The reaction between carbon monoxide and hydrogen is an example of hydrogenation. Liquid oils are often hydrogenated. Hydrogenation changes the liquid oil to a solid fat. Most kitchens contain foods with hydrogenated or partially hydrogenated oils. Vegetable shortening, such as Crisco, is a good example. Hydrogenation makes it easier to pack and transport oils.

Hydrogen is also used in oxyhydrogen ("oxygen + hydrogen") and atomic hydrogen torches. These torches produce temperatures of a few thousand degrees. At these temperatures, it is possible to cut through steel and most other metals. These torches can also be used to weld (join together with heat) two metals.

Another use for hydrogen is in lighter-than-air balloons. Hydrogen is the least dense of all gases. So a balloon filled with hydrogen can lift very large loads. Such balloons are not used to carry people. The danger of fire or explosion is too

Solving the world's energy problems

Most people don't worry about filling their cars with gas. They seem to believe that there will always be enough coal, oil, and natural gas to keep civilization running. Those three fuels—the "fossil fuels"—are what keep people on the move today. They fuel cars and trucks, heat homes and offices, and keep factories operating.

But fossil fuels will not last forever. At some point, all the coal, oil, and natural gas will be gone. What source of energy will humans turn to?

Some people believe that hydrogen is the answer. They talk about the day when the age of fossil fuels will be replaced by a hydrogen economy.

"Hydrogen economy" refers to a world in which the burning of hydrogen will be the main source of energy and power. Hydrogen seems to be a good choice for future energy needs. When it burns, it produces only water:

$$2H_2 + O_2 \rightarrow 2H_2O$$

A lot of energy is produced in this reaction. That energy can be used to operate cars, trucks, trains, boats, and airplanes. It can be used as a source of heat for keeping people warm and running chemical reactions.

Why doesn't a hydrogen economy exist today? The answer is easy. It is still too expensive to make hydrogen gas. No one has found a way to remove hydrogen from water or some other source at a low cost. It is still cheaper to mine for coal or drill for oil than to make hydrogen.

But that may not always be true. Some day, someone will find a way to make hydrogen cheaply. When that happens, the day of the hydrogen economy will have arrived.

great. On May 6, 1937, a hydrogen fire destroyed the German airship *Hindenburg,* as it was landing in Lakehurst, New Jersey; 36 people died. Today, hydrogen balloons are used for lifting weather instruments into the upper atmosphere.

One of the best known uses of hydrogen is as a rocket fuel. Many rockets obtain the power they need for lift-off by burning oxygen and hydrogen in a closed tank. The energy produced by this reaction provides thrust to the rocket.

Compounds

Millions of hydrogen compounds are known. One of the most important groups of hydrogen compounds is the acids. An acid is any compound that contains hydrogen as its positive part. Common acids include: hydrochloric acid (HCl), sulfuric acid (H_2SO_4), nitric acid (HNO_3), acetic acid ($HC_2H_3O_2$), phosphoric acid (H_3PO_4), and hydrofluoric acid (HF).

Acids are present in thousands of natural substances and artificial products. The following list gives a few examples: vinegar, or acetic acid ($HC_2H_3O_2$); sour milk, or lactic acid ($C_3H_6O_3$); lemons and other citrus fruits, or citric acid ($C_6H_8O_7$); soda water, or carbonic acid (H_2CO_3); battery acid, or sulfuric acid (H_2SO_4); and boric acid (H_3BO_3).

Health effects

Hydrogen is essential to every plant and animal. Nearly every compound in a living cell contains hydrogen. It is harmless to humans unless taken in very large amounts. In this case, it is dangerous only because it cuts off the supply of oxygen humans need to breathe.

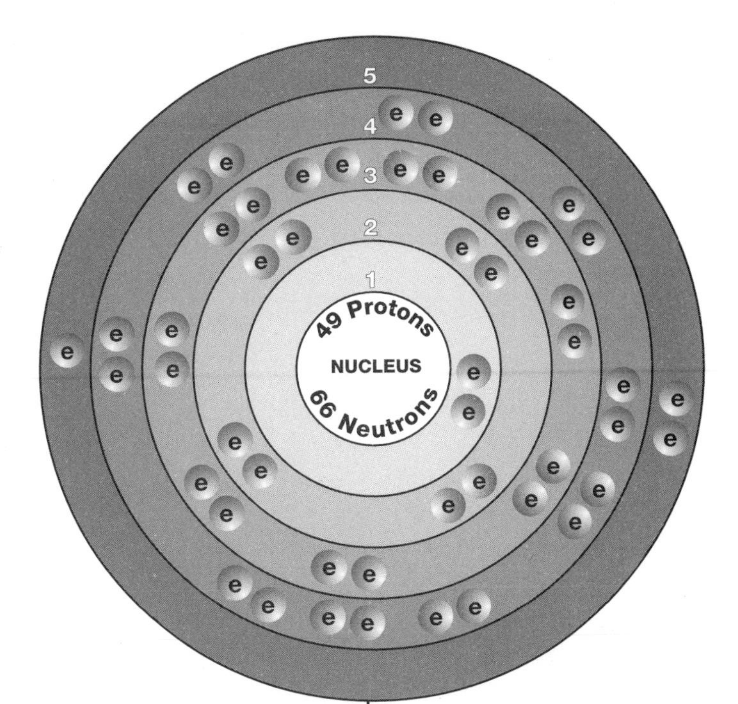

INDIUM

SYMBOL
In

ATOMIC NUMBER
49

ATOMIC MASS
114.82

FAMILY
Group 13 (IIIA)
Aluminum

PRONUNCIATION
IN-dee-um

Overview

Indium is part of the aluminum family in Group 13 (IIIA) of the periodic table. The periodic table is a chart that shows how chemical elements are related to each other. Indium was discovered in 1863 by German chemists Ferdinand Reich (1799–1882) and Hieronymus Theodor Richter (1824–98).

Indium has a number of interesting properties. For example, it has a low melting point for metals, 156.6°C (313.9°F). When pure, it sticks very tightly to itself or to other metals. This property makes it useful as a solder. Solder is a material used to join two metals to each other. Other uses of indium are in the manufacture of batteries, electronic devices, and in research.

Discovery and naming

Between 1860 and 1863, indium, **cesium, rubidium,** and **thallium** were found using spectroscopy. Spectroscopy is the process of analyzing light produced when an element is heated. The light produced is different for every element. The spectrum (plural: spectra) of an element consists of a series of colored lines. Reich and Richter also produced the first impure sample of indium in 1863.

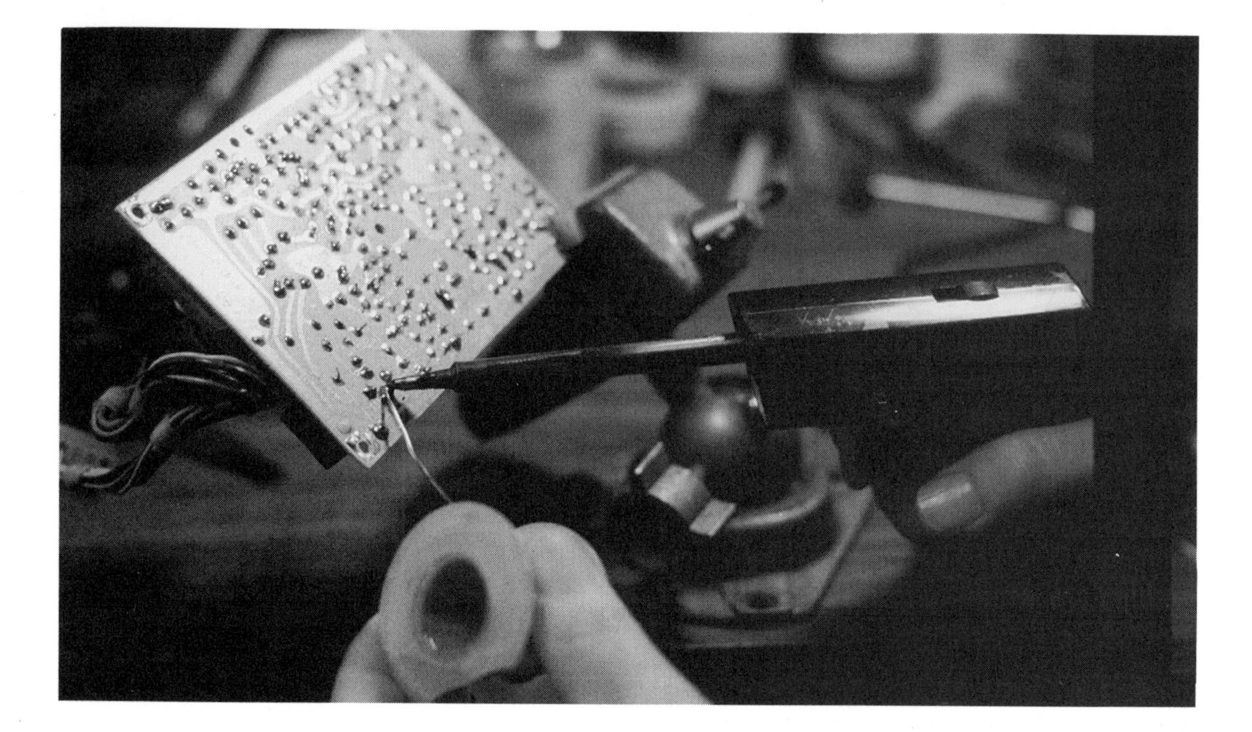

When indium is in its pure form, it sticks very tightly to itself or to other metals, making it useful as a solder. Here, a circuit board is being soldered.

Scientists use a spectroscope in this process. The spectroscope was invented in 1814 by German physicist Joseph von Fraunhofer (1787–1826). Forty years later, German chemists Robert Bunsen (1811–99) and Gustav Robert Kirchhoff (1824–87) improved on the instrument. They showed how it could be used to study the chemical elements.

Reich and Richter suggested the name indium for the element they discovered because its main spectral lines are a brilliant indigo blue.

Indium's main spectral lines are a brilliant indigo blue.

Physical properties

Indium is a silvery-white, shiny metal with a density of 7.31 grams per cubic centimeter. It is one of the softest metals known, even softer than **lead.** If drawn across a piece of paper, it leaves a mark like a "lead" pencil (which is actually carbon).

An unusual property of indium is that it produces a "tin cry." A **tin** cry is a scream-like sound made when the metal is bent.

Indium has a melting point of 156.6°C (313.9°F) and a boiling point of 2,075°C (3,767°F). It has the unusual property of remaining soft and workable at very low temperatures. This property allows it to be used in special equipment needed for temperatures near absolute zero. Absolute zero is the coldest temperature possible. It is about –273°C (–459°F).

Chemical properties

Indium metal dissolves in acids, but does not react with **oxygen** at room temperature. At higher temperatures, it combines with oxygen to form indium oxide (In_2O_3).

Occurrence in nature

Indium is relatively rare. Its abundance in the Earth's crust is estimated to be about 0.1 part per million. That makes it slightly more abundant than **silver** or **mercury.**

Indium is generally found in ores of **zinc.** The metal is not usually produced in the United States. It is imported from Canada, China, and Russia. In 1996, about 45 tons of the metal were used in the United States.

Isotopes

Two naturally occurring isotopes of indium exist, indium-113 and indium-115. Isotopes are two or more forms of an element. Isotopes differ from each other according to their mass number. The number written to the right of the element's name is the mass number. The mass number represents the number of protons plus neutrons in the nucleus of an atom of the element. The number of protons determines the element, but the number of neutrons in the atom of any one element can vary. Each variation is an isotope.

Indium-115 is radioactive. A radioactive isotope is one that breaks apart and gives off some form of radiation. Indium-115 has a half life of about 440 trillion years. The half life of a radioactive element is the time it takes for half of a sample of the element to break down. Starting with 10 grams of indium-115 today, only 5 grams would be left 440 trillion years from now.

A number of artificial radioactive isotopes of indium also exist. Radioactive isotopes are produced when very small particles are fired at atoms. These particles stick in the atoms and make them radioactive. Two of these isotopes are used in medicine. Indium-113 is used to examine the liver, spleen, brain, pulmonary ("breathing") system, and heart and blood system. Indium-111 is used to search for tumors, internal bleeding, abscesses, and infections and to study the gastric (stomach)

WORDS TO KNOW

Absolute zero the lowest temperature possible, about –273°C (–459°F)

Alloy a mixture of two or more metals that has properties different from those of the individual metals

Half life the time it takes for half of a sample of a radioactive element to break down

Isotopes two or more forms of an element that differ from each other according to their mass number

Radioactivity the process by which an isotope or element breaks down and gives off some form of radiation

Solder a material used to join two metals to each other

Spectroscopy the process of analyzing light produced when an element is heated

Indium samples.

and blood systems. In both cases, the radioactive isotope is injected into the blood stream. Inside the body, the isotope gives off radiation. That radiation can be detected by means of a camera or other device. The radiation pattern observed provides information about the organ or system being studied.

Extraction

Indium is obtained in its pure form by separating it from zinc and other elements in zinc ores.

Uses and compounds

The primary use of indium is in making alloys. An alloy is made by melting and mixing two or more metals. The mixture has properties different from those of the individual metals. Indium has been called a "metal vitamin" in alloys. That means that very small amounts of indium can make big changes in an alloy. For example, very small amounts of indium are sometimes added to **gold** and **platinum** alloys to make them much

harder. Such alloys are used in electronic devices and dental materials.

Indium is also added to solders. It reduces the melting point of some solders, strengthens other solders, and prevents still other solders from breaking down too easily.

The single most important use of indium is in making coatings. For example, some aircraft parts are made of alloys that contain indium. The indium prevents them from wearing out or reacting with oxygen in the air.

Alloys and compounds of indium are also used in making optical (light) devices. For example, indium **gallium** arsenide (InGaAs) is able to convert pulses of light into electrical signals. One application of a device like this is in solar cells. A solar cell is a device used to change sunlight into electrical current. Many scientists think that solar cells may replace coal, oil, and natural gas for many purposes in the future.

Very small amounts of indium are sometimes added to gold and platinum alloys to make them much harder.

In 1997, the Siemens Corporation announced the largest and most advanced solar conversion system ever developed. It uses **copper** indium diselinide ($CuInSe_2$) in its solar cells. The system produces one million watts of electrical energy from sunlight. That is enough electricity to completely operate a large office building.

Health effects

The health effects of indium compounds are somewhat unusual. When taken by mouth, they are relatively harmless. When injected into the skin, however, they are very poisonous.

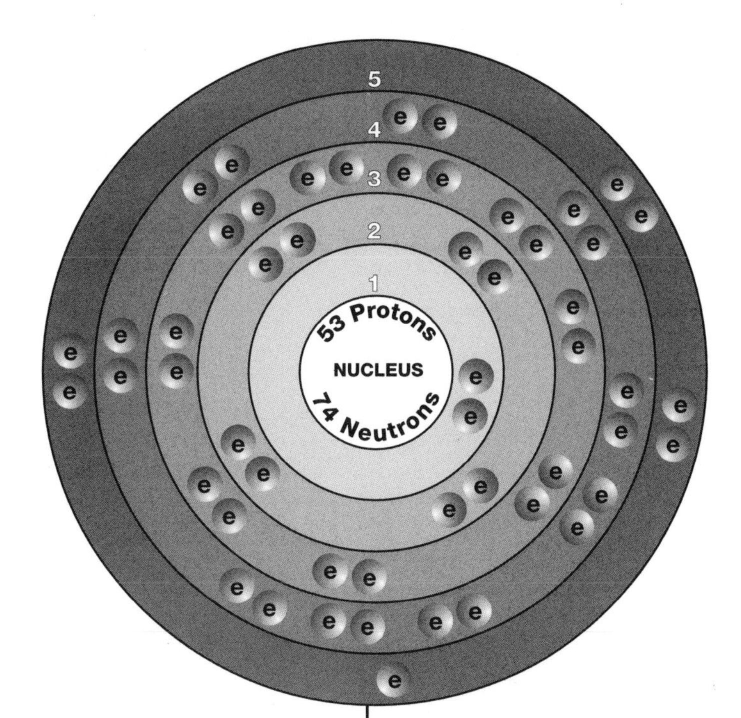

IODINE

Overview

Iodine is the heaviest of the commonly occurring halogens. The halogens are in Group 17 (VIIA) of the periodic table. The periodic table is a chart that shows how chemical elements are related to each other. Iodine's chemical properties are similar to the lighter halogens above it, **fluorine, chlorine,** and **bromine.** But its physical appearance is very different. It is a steel-gray solid that changes into beautiful purple vapors when heated.

Iodine was discovered in 1811 by French chemist Bernard Courtois (1777–1838). The element occurs primarily in seawater and in solids formed when seawater evaporates. Its single most important property may be the ability to kill germs. It is used in antiseptics, germicides (products that kill germs), and other medical applications. However, it has a great many other less common, but important, commercial applications.

SYMBOL
I

ATOMIC NUMBER
53

ATOMIC MASS
126.9045

FAMILY
Group 17 (VIIA)
Halogen

PRONUNCIATION
EYE-uh-dine

Discovery and naming

One of Courtois' first jobs was to assist his father in making compounds of **sodium** and **potassium** from seaweed. Seaweed plants take sodium and potassium compounds out of seawater. The compounds become part of the growing seaweed.

WORDS TO KNOW

Antiseptic a chemical that stops the growth of germs

Catalyst a substance used to speed up or slow down a chemical reaction without undergoing any change

Halogen one of the elements in Group 17 (VIIA) of the periodic table

Isotopes two or more forms of an element that differ from each other according to their mass number

Periodic table a chart that shows how the chemical elements are related to each other

Radioactive isotope an isotope that breaks apart and gives off some form of radiation

Sublimation the process by which a solid turns directly to a gas without first melting

Tincture a solution made by dissolving a substance in alcohol

Courtois and his father collected seaweed on the coasts of Normandy and Brittany in France. Then they burned it. Next, they soaked the seaweed ashes in water to dissolve the sodium and potassium compounds. Sulfuric acid was added to react with the unwanted seaweed chemicals. Finally, they allowed the water to evaporate, leaving the compounds behind. These compounds are white crystals, much like ordinary table salt. The compounds were sold to large industrial businesses for use in such products as table salt and baking soda.

One day in 1811, Courtois made a mistake. He added too much sulfuric acid to the mixture. He was amazed to see clouds of beautiful violet vapor rising from the mixture. He decided to study the new material. Eventually, he proved it was a new element. He named the element after its color. In Greek, the word *iodes* means "violet."

Physical properties

Iodine is one of the most striking and beautiful of all elements. As a solid, it is a heavy, grayish-black, metallic-looking material. When heated, it does not melt. Instead, it sublimes. Sublimation is the process by which a solid turns directly to a gas without first melting. The resulting iodine vapor has a violet color and a harsh odor. If a cold object, such as an **iron** bar, is placed in these vapors, iodine changes back to a solid. It forms attractive, delicate, metallic crystals.

Iodine dissolves only slightly in water. But it dissolves in many other liquids to give distinctive purple solutions. If heated under the proper conditions, it can be made to melt at 113.5°C (236.3°F) and to boil at 184°C (363°F). The density of the element is 4.98 grams per cubic centimeter.

Chemical properties

Like the other halogens, iodine is an active element. However, it is less active than the three halogens above it in the periodic table. Its most common compounds are those of the alkali metals, sodium, and potassium. But it also forms compounds with other elements. It even forms compounds with the other halogens. Some examples are iodine monobromide (IBr), iodine monochloride (ICl), and iodine pentafluoride (IF_5).

A magnified view of a crystal of iodine.

Occurrence in nature

Iodine is not very abundant in the Earth's crust. Its abundance is estimated to be about 0.3 to 0.5 parts per million. It ranks in the bottom third of the elements in terms of abundance. It is still more common than **cadmium, silver, mercury,** and **gold.** Its abundance in seawater is estimated to be even less, about 0.0003 parts per million.

Iodine tends to be concentrated in the Earth's crust in only a few places. These places were once covered by oceans. Over millions of years, the oceans evaporated. They left behind the chemical compounds that had been dissolved in them. The dry chemicals left behind were later buried by earth movements. Today, they exist underground as salt mines.

Iodine can also be collected from seawater, brackish water, brine, or sea kelp. Seawater is given different names depending on the amount of solids dissolved in it. Brackish water, for

A mistake by Bernard Courtois led to clouds of beautiful violet vapor rising from a mixture on which he was working. It was iodine.

Kelp, a type of seaweed, is a popular source of iodine, since it absorbs the element from seawater.

example, has a relatively low percentage of solids dissolved in water. The range that is sometimes given is 0.05 to 3 percent solids in the water. Brine has a higher percentage of dissolved solids. It may contain anywhere from 3 to 20 percent of solids dissolved in water.

Finally, sea kelp is a form of seaweed. As it grows, it takes iodine out of seawater. Over time, sea kelp has a much higher concentration of iodine than seawater. Sea kelp is harvested, dried, and burned to collect iodine. The process is not much different from the one used by Courtois in 1811.

Iodine compounds are used in the production of photographic film.

Isotopes

Only one naturally occurring isotope of iodine is known, iodine-127. Isotopes are two or more forms of an element. Isotopes differ from each other according to their mass number. The number written to the right of the element's name is the mass number. The mass number represents the number of protons plus neutrons in the nucleus of an atom of the element. The number of protons determines the element, but the number of neutrons in the atom of any one element can vary. Each variation is an isotope.

Approximately 30 radioactive isotopes of iodine have been made artificially. A radioactive isotope is one that breaks apart and gives off some form of radiation. Radioactive isotopes are produced when very small particles are fired at atoms. These particles stick in the atoms and make them radioactive.

A number of iodine isotopes are used commercially. In medical applications, these isotopes are injected into the body or given to the patient through the mouth. The isotopes then travel through the body in the bloodstream. As they travel, they give off radiation. That radiation can be detected by using X-ray film. A medical specialist can tell how well the body is functioning by observing the pattern of radiation.

Iodine and human health

The amount of iodine in the human body is very small. To find out how much is in one's body, one's body weight is divided by 2,500,000. That number is the weight of iodine in the body. For normal people, the amount is about equal to the size of the head of a pin.

That tiny dot of iodine can mean the difference between good and bad health. People who do not have enough can develop serious health problems. At one time, the most common of those problems was a disease known as goiter. Goiter causes a large lump in the neck as the thyroid grows out of control. (It can grow as large as a grapefruit.) A goiter tries to make thyroid hormones, but it does not receive enough iodine from the person's diet. So it keeps expanding, trying to do its job.

A lack of iodine can cause other problems too. For example, thyroid hormones are needed for normal brain development in an unborn child. They are also needed to continue that development after birth. People who do not include enough iodine in their diet do not develop normally. Today, experts say that low levels of iodine are the leading cause of mental retardation, deafness, mutism (the inability to speak), and paralysis. They also say less serious problems can be blamed on low iodine levels. These include lethargy drowsiness, clumsiness, and learning disabilities.

Low iodine levels can be easily corrected. In most developed countries today, companies that make table salt add a small amount of potassium iodide (KI) to their salt. The salt is labeled "iodized salt." People who use it get all the iodine they need for normal thyroid function.

But people who live in developing countries may not be able to get iodized salt. The World Health Organization (WHO) is trying to find ways of providing iodine to these people. The WHO estimates that 1.5 billion people live in areas where levels of iodine are low. Up to 20 million of these people may have mental disabilities because of a lack of iodine. The WHO has started a program to ensure that future generations in these regions get the iodine needed to develop and function normally.

Iodine isotopes are used in many ways. Iodine-123 is used in studies of the brain, kidneys, and thyroid. Iodine-125 is used in studies of the pancreas, blood flow, thyroid, liver, take-up of minerals in bones, and loss of proteins in the body. And iodine-131 is used in studies of the liver, kidneys, blood flow, lungs, brain, pancreas, and thyroid.

The most common iodine isotope used is iodine-131. When iodine (of any kind) enters the body, it tends to go directly to the thyroid. The iodine is then used to make thyroid hormones. If radioactive iodine is used, a doctor can tell how well the thyroid gland is working. If a high level of radiation is given off, the gland may be overactive. If a low level of radiation is given off, the gland may be underactive. In either case, the person may need some treatment to help the thyroid gland work more normally.

Extraction

When a mixture of substances containing iodine is heated, the iodine sublimes. It can then be collected and purified.

Uses and compounds

About two-thirds of all iodine and its compounds are used in sanitation systems or in making various antiseptics and drugs. Iodine is also used to make dyes, photographic film, and specialized soaps. It is used in some industries as a catalyst. A catalyst is a substance used to speed up or slow down a chemical reaction. The catalyst does not undergo any change itself during the reaction.

Iodine kills bacteria and other disease-causing organisms.

Health effects

Iodine can have both favorable and unfavorable effects on living organisms. It tends to kill bacteria and other disease-causing organisms. In fact, this property leads to its use in sanitation systems and as an antiseptic. An antiseptic is a chemical that stops the growth of germs. Not so long ago, tincture of iodine was one of the most popular antiseptics. It was applied to cuts and wounds to prevent infection. Tincture is a solution made by dissolving some substance (such as iodine) in alcohol rather than in water. Today, tincture of iodine has been replaced by other antiseptics.

One reason that tincture of iodine is used less today is that it can also cause problems. In higher doses, iodine can irritate or burn the skin. It can also be quite poisonous if taken internally.

Iodine plays an important role in the health of plants and animals. It is needed to maintain good health and normal growth. In humans, iodine is used to make a group of important compounds known as thyroid hormones. These chemicals are produced in the thyroid gland at the base of the neck. These chemicals control many important bodily functions. A lack of thyroid hormones can result in the disorder known as goiter. Goiter causes a large lump in the neck as the thyroid grows out of control. Iodine is added to table salt today, so goiter is rarely seen in the United States.

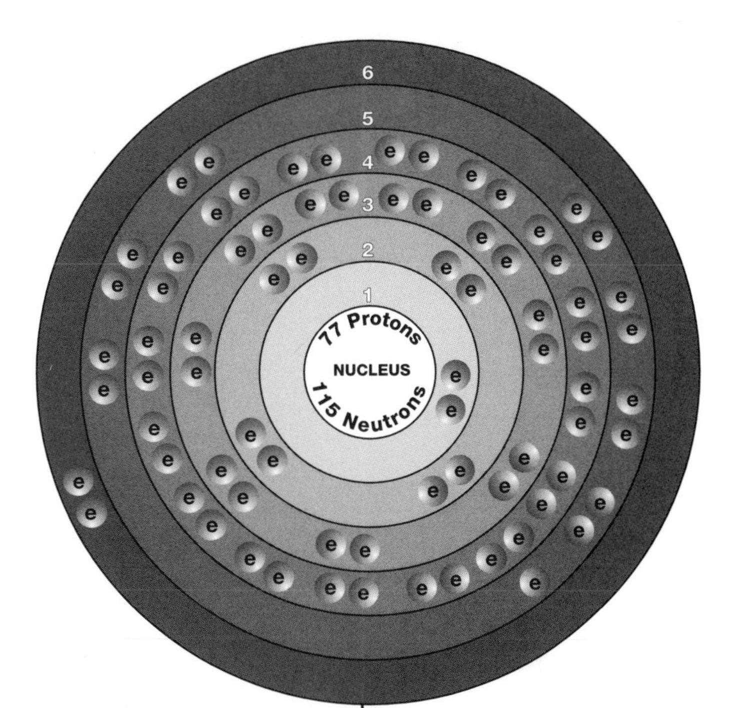

IRIDIUM

SYMBOL
Ir

ATOMIC NUMBER
77

ATOMIC MASS
192.217

FAMILY
Group 9 (VIIIB)
Transition metal
Platinum group

PRONUNCIATION
i-RI-dee-um

Overview

Iridium is in Group 9 (VIIIB) of the periodic table. The periodic table is a chart that shows how elements are related to one another. Iridium is a transition metal that is also part of the platinum family.

The metals in the platinum family are also known as the noble metals. They have this name because they do not react well with other elements and compounds. They appear to be "too superior" to react with most other substances. In fact, iridium is the most corrosion-resistant metal known. It is not affected by high temperatures, acids, bases, or most other strong chemicals. That property makes it useful in making objects that are exposed to such materials.

Iridium may be a key element in the puzzle of dinosaur extinction. Scientists search for iridium in the soil to track the impact of a giant meteor with the Earth 65 million years ago.

Discovery and naming

The platinum metals posed a difficult problem for early chemists. These metals often occurred mixed together in the earth. When a scientist thought that he was analyzing a sam-

WORDS TO KNOW

Alloy a mixture of two or more metals with properties different from those of any one metal alone

Aqua regia a mixture of two strong acids—nitric acid and hydrochloric acid

Density the mass of a substance per unit volume

Ductile capable of being drawn into thin wires

Halogens elements in Group 17 (VIIA) of the periodic table

Isotopes two or more forms of an element that differ from each other according to their mass number

Noble metals *see* **Platinum family**

Platinum family a group of elements that occur close to platinum in the periodic table and with platinum in the Earth's surface

Radioactive isotope an isotope that gives off radiation and changes into a new form

Reactive combines with other substances relatively easily

ple of platinum, the sample often contained iridium, **rhodium,** **osmium,** and other metals as well. The work of French chemist Pierre-François Chabaneau is an example. In the late 1780s, the Spanish government gave its entire supply of platinum to Chabaneau to study. But Chabaneau's experiments puzzled him. Sometimes the platinum he worked with could be hammered into flat plates easily. At other times, it was brittle and shattered when hammered. Chabaneau did not realize that the "platinum" he was studying included various amounts of other noble metals.

In the early 1800s, a number of chemists worked to separate the platinum metals. One of those chemists was an Englishman named Smithson Tennant (1761–1815). Like so many others, Tennant became interested in chemistry at an early age. He is said to have made gunpowder to use in fireworks when he was only nine years old!

In 1803, Tennant attempted to dissolve platinum in aqua regia. Aqua regia is a mixture of two strong acids—nitric acid and hydrochloric acid. He found that most of the platinum metal dissolved, leaving a small amount of black powder. Other chemists had not bothered to study the powder. But Tennant did. He discovered that it had properties very different from those of platinum. He realized he had discovered a new element. He named it iridium, from the Greek goddess Iris, whose symbol is a rainbow. Tennant chose this name because the compounds of iridium have so many different colors. For example, iridium potassium chloride (K_2IrCl_6) is dark red, iridium tribromide ($IrBr_3$) is olive-green, and iridium trichloride ($IrCl_3$) is dark green to blue-black.

Physical properties

Iridium metal is silvery-white with a density of 22.65 grams per cubic centimeter. A cubic centimeter of iridium weighs 22.65 times as much as a cubic centimeter of water. It is the most dense element known. Iridium has a melting point of 2,443°C (4,429°F) and a boiling point of about 4,500°C (8,130°F). Cold iridium metal cannot be worked easily. It tends to break rather than bend. It becomes more ductile (flexible) when hot. Ductile means capable of being drawn into thin wires. Therefore, it is usually shaped at high temperatures.

Small parts of iridium can be found in meteorites. The Barrington Crater, in northern Arizona, was created about 25,000 years ago by a meteorite the size of a large house. It hit the ground at 9 miles per second, and created a hole .7 miles (1.2 kilometers) across and 590 feet (180 meters) deep.

Chemical properties

Iridium is unreactive at room temperatures. When exposed to air, it reacts with oxygen to form a thin layer of iridium dioxide (IrO_2).

$$Ir + O_2 \rightarrow IrO_2$$

At high temperatures, the metal becomes more reactive. Then it reacts with oxygen and halogens to form iridium dioxide and iridium trihalides. For example:

$$2Ir + 3Cl_2 \rightarrow 2IrCl_3$$

Occurrence in nature

Iridium is one of the rarest elements in the Earth's crust. It is thought to exist in two parts per billion. Interestingly, it is more abundant in other parts of the universe. Iron meteorites, for example, generally contain about 3 parts per million of irid-

Tracking the fate of the dinosaurs

Why did the dinosaurs die out? This question has long been one of the most interesting and puzzling issues in science. What happened to make these huge reptiles disappear in such a short period of geological time?

One answer might be found in the Asteroid Disaster Theory. According to this theory, a huge asteroid struck the Earth's surface about 65 million years ago. The exploding asteroid threw enormous amounts of dust into the air. The dust blocked out sunlight for more than a year. Plants on the Earth's surface died. Dinosaurs who lived on those plants died out. So did the meat-eating dinosaurs who lived off the plant eaters.

But how is it possible to know if an asteroid really did hit the Earth's surface 65 million years ago? Scientists have now found an answer. In some parts of the Earth, they have found a layer of the Earth's crust that contains an unusually high level of iridium metal. Iridium rarely occurs on Earth. But it occurs much more commonly in meteors and asteroids. Scientists believe the iridium-rich layer was formed when an asteroid struck the Earth's surface. They believe the event occurred 65 million years ago. This "iridium clue" is a key, therefore, to understanding how dinosaurs disappeared from the Earth.

Iridium is one of the rarest elements in the Earth's crust. But it is more abundant in other parts of the universe. Iridium is found in meteorites.

ium. Stony meteorites contain less iridium, about 0.64 parts per million.

Iridium usually occurs in combination with one or more other noble metals. Two common examples are osmiridium and iridosmine, combinations of iridium and osmium. The most important sources of iridium metal are Canada, South Africa, Russia, and the state of Alaska.

Isotopes

Two naturally occurring isotopes of iridium exist, iridium-191 and iridium-193. Isotopes are two or more forms of an element. Isotopes differ from each other according to their mass number. The number written to the right of the element's name is the mass number. The mass number represents the number of protons plus neutrons in the nucleus of an atom of the element. The number of protons determines the element, but the

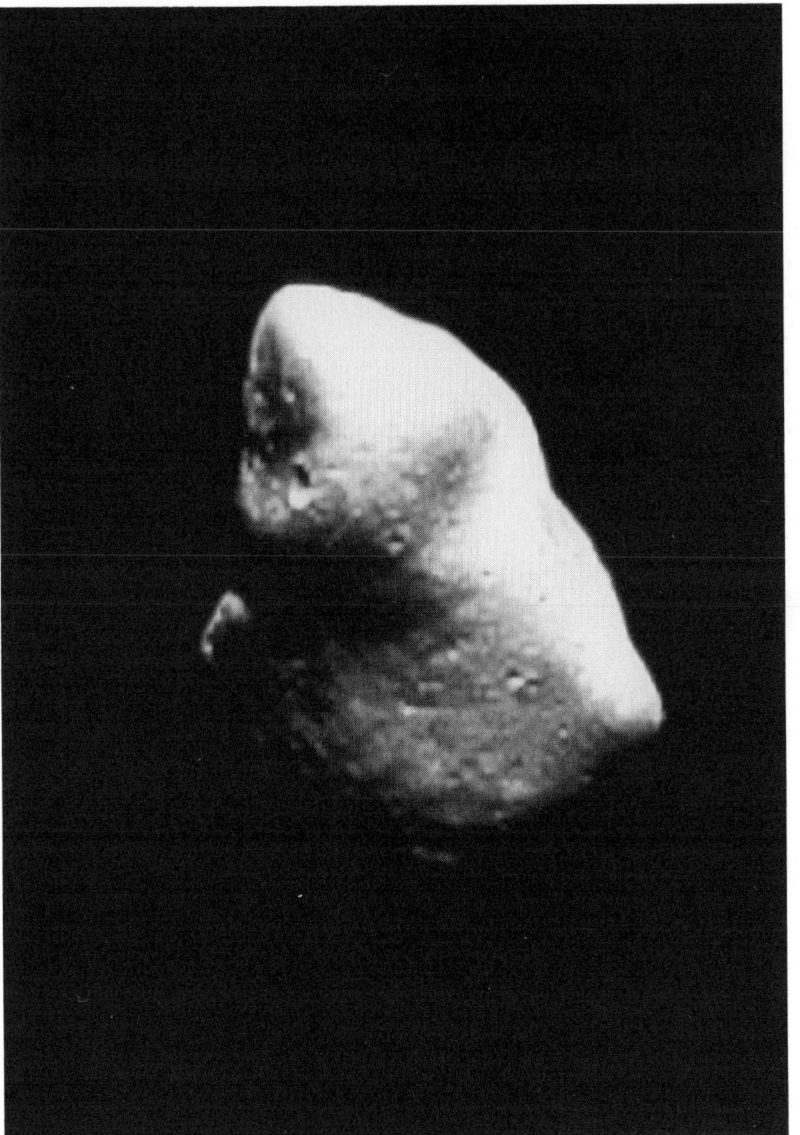

Did an asteroid—which contains iridium—kill off dinosaurs 65 million years ago? The asteroid Gaspra, photographed from the Galileo *space probe, is shown here.*

number of neutrons in the atom of any one element can vary. Each variation is an isotope.

About ten radioactive isotopes of iridium exist. A radioactive isotope is one that gives off radiation and changes into a new form. The only important radioactive isotope of iridium is iridium-192. This isotope has a half-life of 74 days. A half-life is the time it takes for one half of a sample to break down. Iridium-192 is used to make X-ray photographs of metal castings and to treat cancer.

The standard kilogram

Butter comes in one-pound or one-kilogram packages. But who decides how much "one pound" or "one kilogram" of butter is?

Every nation has a governmental office for weights and measures. The office maintains an "official" pound or kilogram. It is usually a piece of metal known to weigh exactly one pound or one kilogram. But how does each nation know exactly what size its official weight should be?

The official world standard for the kilogram is kept at the International Bureau of Weights and Measures in Paris. The standard is a piece of platinum-iridium metal stored in an airtight jar. The standard is made of platinum and iridium to protect it from reacting with oxygen and other chemicals in the air. In this way, the standard's weight will always remain exactly the same.

Extraction

Iridium and the other platinum metals tend to occur together. A series of chemical reactions is used to separate one metal from the other. The other metals are then removed by other techniques. Very little iridium is produced each year, probably no more than a few metric tons.

Uses

One kind of iridium catalyst is able to capture sunlight and turn it into chemical energy.

The primary use of iridium is in the manufacture of alloys. An alloy is made by melting and mixing two or more metals. An alloy's properties differ from those of the elements that make it up. Iridium is often combined with platinum, for example, to provide a stronger material than the platinum itself. These alloys are very expensive and are used for only special purposes. For instance, the sparkplugs used in helicopters are made of a platinum-iridium alloy. Such alloys are also used for electrical contacts, special types of electrical wires, and electrodes.

Iridium metal is increasingly being used as catalysts. Catalysts are substances that speed up a reaction without changing themselves. Iridium catalysts have been used in amazing new products. For example, one kind of iridium catalyst is able to capture sunlight and turn it into chemical energy. That process is similar to the one used by plants in photosynthesis. Finding a synthetic (artificial) way to make photosynthesis happen is one of the great goals of modern chemistry.

The heating of lead ores is called roasting, or smelting. The process results in pure lead.

Niobium alloys are used in superconducting magnets. A superconductor has no resistance to an electric current. Here, a small magnet levitates over a cooled slab of superconducting ceramic.

Neodymium is used in making lasers. This laser beam is reflected off mirrors and through filters.

Magnified view of a lanthanum aluminate crystal.

The glass in satellites often contains germanium. This satellite was launched in June 1990.

Three Mile Island nuclear reactor in Middletown, Pennsylvania. This was the site of a partial meltdown in 1979.

HORSESHOE
FREMONT
GAMBLI

The neon lights of Las Vegas, Nevada, in the early 1990s.

Stars use hydrogen as a fuel with which to produce energy.
Antares—the brightest star in the constellation Scorpius—is shown here.

Gallium melts when held in the hand.

A farmer sprays nitrogen fertilizer on his rice field in California.

At high temperatures, magnesium burns with a blinding white light.

Space technology often uses alloys that are too expensive for everyday use. An example is the propulsion systems used for keeping satellites in place. Some of these systems use alloys made of iridium and another platinum metal, **rhenium.** These alloys remain strong at high temperatures and are not attacked by fuels used in the systems.

Compounds

The compounds of iridium have almost no practical applications. A few are used in coloring ceramics because of their striking colors.

Health effects

Scientists are not aware of any health benefits or risks associated with iridium.

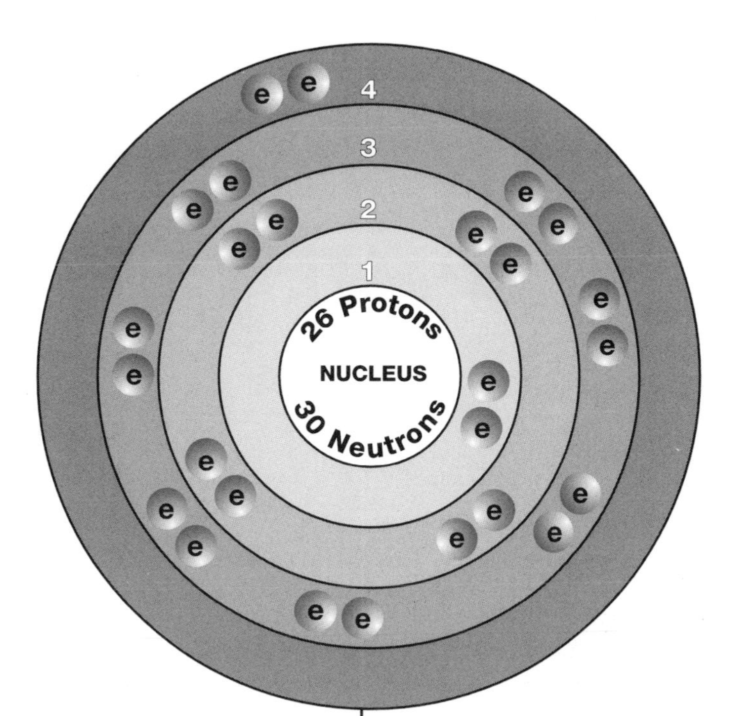

IRON

Overview

The period in human history beginning in about 1200 B.C. is called the Iron Age. It was at about this time that humans first learned how to use iron metal. But in some ways, one could refer to the current era as the New Iron Age. Iron is probably the most widely used and most important metal today. No other metal is available to replace iron in all its many applications.

Iron is a transition metal. The transition metals are the elements that make up Groups 3 through 12 in the periodic table. The periodic table is a chart that shows how elements are related to one another. The transition metals are typical metals in that they tend to be bright, shiny, silvery solids. They all tend to conduct heat and electricity well. And they usually have high melting points.

Iron normally does not occur as a free element in the earth. In fact, iron was not of much value to humans until they learned how to free iron from its compounds. Once they could do that, humans were able to make tools, weapons, household implements, and other objects out of iron. This step marked the beginning of the Iron Age.

SYMBOL
Fe

ATOMIC NUMBER
26

ATOMIC MASS
55.847

FAMILY
Group 8 (VIIIB)
Transition metal

PRONUNCIATION
EYE-urn

Some meteorites are very rich in iron. Here, children play on the Williamette meteorite in Hayden Planetarium in New York City, in 1939.

Iron is most valuable not as a pure metal, but in alloys. An alloy is made by melting and mixing two or more metals. The mixture has properties different from those of the individual metals. The best known and most widely used alloy of iron is steel. Steel contains iron and at least one other element. Today, specialized steels of all kinds are available for many different applications.

Iron is probably the most widely used and most important metal today.

Discovery and naming

Ancient Egyptians had learned how to use iron before the First Dynasty, which began in about 3400 B.C. The Egyptians probably found the iron in meteorites. Meteorites are chunks of rock and metal that fall from the sky. Some meteorites are very rich in iron. The Egyptians made tools and jewelry out of iron.

Iron was also known to early Asian civilizations. In Delhi, India, for example, a pillar made out of iron built in A.D. 415

still stands. It weighs 6.5 metric tons and remains in good condition after nearly 1,600 years.

Early Chinese civilizations also knew about iron. Workers learned to produce iron as early as 200 B.C. A number of iron objects, including cannons, remain from the Han period (202 B.C. to A.D. 221).

The Bible also includes many mentions of iron. For example, a long passage in the book of Job describes the mining of iron. Other passages tell about the processing of iron ore to obtain iron metal.

By the time of the Roman civilization, iron had become an essential metal. The historian Pliny (A.D. 23–79) described the role of iron in Rome:

> It is by the aid of iron that we construct houses, cleave rocks, and perform so many other useful offices of life. But it is with iron also that wars, murders, and robberies are effected, and this, not only hand to hand, but from a distance even, by the aid of weapons and winged weapons, now launched from engines, now hurled by the human arm, and now furnished with feathery wings.

Ancient Egyptians had learned how to use iron before the First Dynasty, which began in about 3400 B.C.

Even from the earliest days, humans probably seldom used iron in a pure form. It was difficult to make iron that was free of impurities, such as **carbon** (charcoal) and other metals. More important, however, it became obvious that iron *with* impurities was a stronger metal that iron *without* impurities.

It was not until 1786, however, that scientists learned what it was in steel that made it a more useful metal than iron. Three researchers, Gaspard Monge (1746–1818), C. A. Vandermonde, and Claude Louis Berthollet (1748–1822) solved the puzzle. They found that a small amount of carbon mixed with iron produced a strong alloy. That alloy was steel. Today, the vast amount of iron used in so many applications is used in the form of steel, not pure iron.

The chemical symbol for iron is Fe. That symbol comes from the Latin name for iron, *ferrum*.

WORDS TO KNOW

Alloy a mixture of two or more metals with properties different from those of the individual metals

Catalyst a substance used to speed up or slow down a chemical reaction without undergoing any change itself

Ductile capable of being drawn into thin wires

Isotopes two or more forms of an element that differ from each other according to their mass number

Malleable capable of being hammered into thin sheets

Periodic table a chart that shows how the chemical elements are related to each other

Radioactive isotope an isotope that breaks apart and gives off some form of radiation

Tensile capable of being stretched without breaking

Tracer a radioactive isotope whose presence in a system can easily be detected

Transition metal an element in Groups 3 through 12 of the periodic table

Workability the ability to work with a metal to get it into a desired shape or thickness

Physical properties

Iron is a silvery-white or grayish metal. It is ductile and malleable. Ductile means capable of being drawn into thin wires. Malleable means capable of being hammered into thin sheets. It is one of only three naturally occurring magnetic elements. The other two are **nickel** and **cobalt.**

Iron has a very high tensile strength. Tensile means it can be stretched without breaking. Iron is also very workable. Workability is the ability to bend, roll, hammer, cut, shape, form, and otherwise work with a metal to get it into a desired shape or thickness.

The melting point of pure iron is 1,536°C (2,797°F) and its boiling point is about 3,000°C (5,400°F). Its density is 7.87 grams per cubic centimeter. The melting point, boiling point, and other physical properties of steel alloys may be quite different from those of pure iron.

Chemical properties

Iron is a very active metal. It readily combines with **oxygen** in moist air. The product of this reaction, iron oxide (Fe_2O_3), is known as rust. Iron also reacts with very hot water and steam to produce **hydrogen** gas. It also dissolves in most acids and reacts with many other elements.

Occurrence in nature

Iron is the fourth most abundant element in the Earth's crust. Its abundance is estimated to be about 5 percent. Most scientists believe that the Earth's core consists largely of iron. Iron is also found in the Sun, asteroids, and stars outside the solar system.

The most common ores of iron are hematite, or ferric oxide (Fe_2O_3); limonite, or ferric oxide (Fe_2O_3); magnetite, or iron oxide (Fe_3O_4); and siderite, or iron carbonate ($FeCO_3$). An increasingly important source of iron is taconite. Taconite is a mixture of hematite and silica (sand). It contains about 25 percent iron.

The largest iron resources in the world are in China, Russia, Brazil, Canada, Australia, and India. The largest producers of iron from ore in the world are China, Japan, the United States, Russia, Germany, and Brazil.

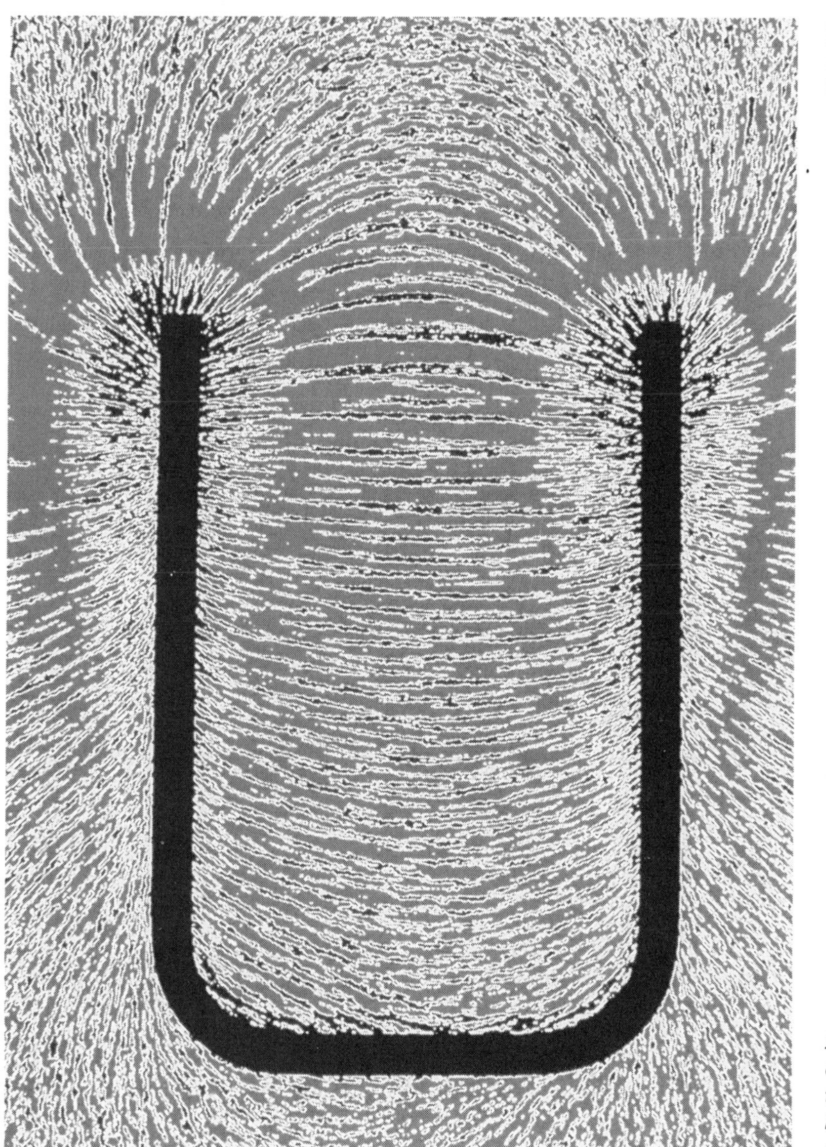

Iron is one of only three naturally occurring magnetic elements. This is a computer graphic of a horseshoe magnet with iron filings aligned around it.

Isotopes

There are four naturally occurring isotopes of iron, iron-54, iron-56, iron-57, and iron-58. Isotopes are two or more forms of an element. Isotopes differ from each other according to their mass number. The number written to the right of the element's name is the mass number. The mass number represents the number of protons plus neutrons in the nucleus of an atom of the element. The number of protons determines the element, but the number of neutrons in the atom of any one element can vary. Each variation is an isotope.

Six radioactive isotopes of iron are known also. A radioactive isotope is one that breaks apart and gives off some form of radiation. Radioactive isotopes are produced when very small particles are fired at atoms. These particles stick in the atoms and make them radioactive.

Two radioactive isotopes of iron are used in medical and scientific research. They are iron-55 and iron-59. These isotopes are used primarily as tracers in studies on blood. A tracer is a radioactive isotope whose presence in a system can easily be detected. The isotope is injected into the system. Inside the system, the isotope gives off radiation. That radiation can be followed by detectors placed around the system. Iron-55 and iron-59 are used to study the way in which red blood cells develop in the body. These studies can be used to tell if a person's blood is healthy.

Most scientists believe that the Earth's core consists largely of iron.

Extraction

Iron goes through a number of stages between ore and final steel product. In the first stage, iron ore is heated with limestone and coke (pure carbon) in a blast furnace. A blast furnace is a very large oven in which the temperature may reach 1,500°C (2,700°F). In the blast furnace, coke removes oxygen from iron ore:

$$3C + 2Fe_2O_3 \xrightarrow{\text{heat}} 3CO_2 + 4Fe$$

The limestone removes impurities in the iron ore.

Iron produced by this method is about 91 to 92 percent pure. The main impurity left is carbon from the coke used in the furnace. This form of iron is known as pig iron. Pig iron is generally too brittle (it breaks too easily) to be used in most products.

A number of methods have been developed for purifying pig iron. A common method used today is called the basic oxygen process. In this process, pig iron is melted in a large oven. Then pure oxygen gas is blown through the molten pig iron. The oxygen burns off much of the carbon in the pig iron:

$$C + O_2 \xrightarrow{\text{heat}} CO_2$$

Although now outdated, iron stoves were once the primary source of heat for homes, as well as a means for cooking.

A small amount of carbon remains in the iron. The iron produced in this reaction is known as steel.

The term "steel" actually refers to a wide variety of products. The various forms of steel all contain iron and carbon. They also contain one or more other elements, such as **silicon, titanium, vanadium, chromium, manganese,** cobalt, nickel, **zirconium, molybdenum,** and **tungsten.** Two other steel-like products are cast iron and wrought iron. Cast iron is an alloy of iron, carbon, and silicon. Wrought iron contains iron and any

one or more of many other elements. In general, however, wrought iron tends to contain very little carbon.

Uses

It would be impossible to list all uses of iron and steel products. In general, those products can be classified into categories: (1) automotive; (2) construction; (3) containers, packaging, and shipping; (4) machinery and industrial equipment; (5) rail transportation; (6) oil and gas industries; (7) electrical equipment; and (8) appliances and utensils. (For more information on specific kinds of steel alloys, see individual elements, such as titanium, vanadium, chromium, manganese, molybdenum, and tungsten.)

The U.S. Recommended Daily Allowance (USRDA) for iron is 18 milligrams.

Compounds

Some iron is made into compounds. The amount is very small compared to the amount used in steel and other iron alloys. Probably the fastest growing use of iron compounds is in water treatment systems. The terms ferric and ferrous refer to two different forms in which iron occurs in compounds. Some of the important iron compounds are:

ferric acetate ($Fe(C_2H_3O_2)_3$): used in the dyeing of cloth

ferric ammonium oxalate ($Fe(NH_4)_3(C_2O_4)_4$): blueprints

ferric arsenate ($FeAsO_4$): insecticide

ferric chloride ($FeCl_3$): water purification and sewage treatment systems; dyeing of cloth; coloring agent in paints; additive for animal feed; etching material for engraving, photography, and printed circuits

ferric chromate ($Fe_2(CrO_4)_3$): yellow pigment (coloring) for paints and ceramics

ferric hydroxide ($Fe(OH)_3$): brown pigment for coloring rubber; water purification systems

ferric phosphate ($FePO_4$): fertilizer; additive for animal and human foods

ferrous acetate ($Fe(C_2H_3O_2)_2$): dyeing of fabrics and leather; wood preservative

ferrous gluconate ($Fe(C_6H_{11}O_7)_2$): dietary supplement in "iron pills"

ferrous oxalate (FeC_2O_4): yellow pigment for paints, plastics, glass, and ceramics; photographic developer

ferrous sulfate ($FeSO_4$): water purification and sewage treatment systems; catalyst in production of ammonia; fertilizer; herbicide; additive for animal feed; wood preservative; additive to flour to increase iron levels

Health effects

Iron is of critical importance to plants, humans, and animals. It occurs in hemoglobin, a molecule that carries oxygen in the blood. Hemoglobin picks up oxygen in the lungs, and carries it to the cells. In the cells, oxygen is used to produce energy the body needs to survive, grow, and stay healthy.

The U.S. Recommended Daily Allowance (USRDA) for iron is 18 milligrams. The USRDA is the amount of an element that a person needs to stay healthy. Iron is available in a number of foods, including meat, eggs, and raisins.

An iron deficiency (lack of iron) can cause serious health problems in humans. For instance, hemoglobin molecules may not form in sufficient numbers. Or they may lose the ability to carry oxygen. If this occurs, a person develops a condition known as anemia. Anemia results in fatigue. Severe anemia can result in a lowered resistance to disease and an increase in heart and respiratory (breathing) problems. Some forms of anemia can even cause death.

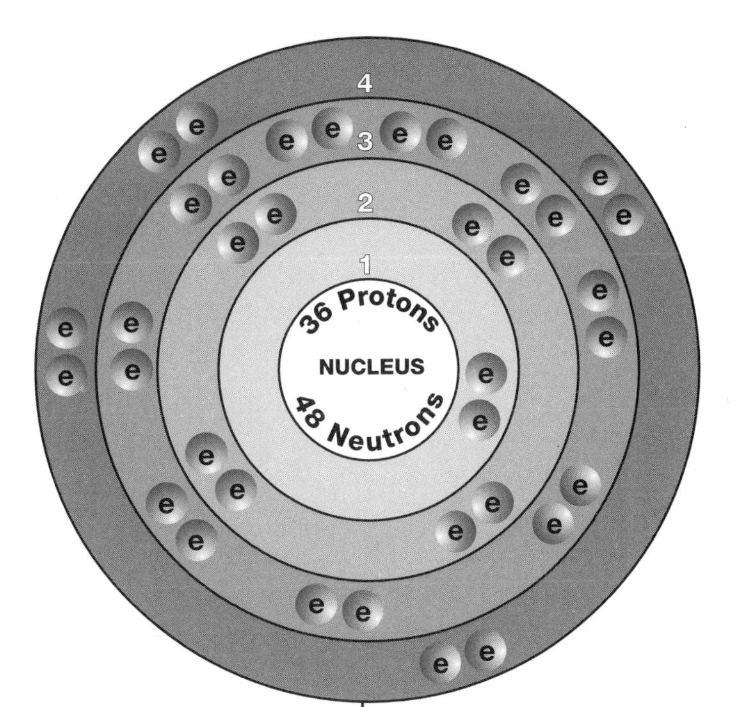

KRYPTON

Overview

Krypton was one of three noble gases discovered in 1898 by Scottish chemist and physicist Sir William Ramsay (1852–1916) and English chemist Morris William Travers (1872–1961). Ramsay and Travers discovered the gases by allowing liquid air to evaporate. As it did so, each of the gases that make up normal air boiled off, one at a time. Three of those gases—krypton, **xenon,** and **neon,** were discovered for the first time this way.

The term noble gas refers to elements in Group 18 (VIIIA) of the periodic table. The periodic table is a chart that shows how chemical elements are related to each other. These gases have been given the name "noble" because they act as if they are "too arrogant" to react with other elements. Until the 1960s, no compound of these gases was known. Since they are so inactive, they are also called the inert gases. Inert means inactive.

Krypton has relatively few commercial uses. All of them involve lighting systems in one way or another.

SYMBOL
Kr

ATOMIC NUMBER
36

ATOMIC MASS
83.80

FAMILY
Group 18 (VIIIA)
Noble gas

PRONUNCIATION
KRIP-ton

Discovery and naming

By 1898, two members of the noble gas family had been discovered. They were **helium** (atomic number 2) and **argon** (atomic number 18). But no other elements in the family had been found. The periodic table contained empty boxes between helium and argon and below argon. The missing noble gases had atomic numbers 10, 36, 54, and 86. Chemists think of empty boxes in the periodic table as "elements waiting to be discovered."

Since the two known noble elements, helium and argon, are both gases, Ramsay and Travers hoped the missing elements were also gases. And if they were, they might be found in air. The problem was that air had already been carefully analyzed and found to be about 99.95 percent **oxygen, nitrogen,** and argon. Was it possible that the missing gases were in the last 0.05 percent of air?

To answer the question, the chemists worked not with air itself, but with liquid air. Air becomes liquid simply by cooling it far enough. The colder air becomes, the more gases within it turn into liquids. At −182.96°C (−297.33°F), oxygen changes from a gas into a liquid. At −195.79°C (−320.42°F), nitrogen changes from a gas into a liquid. And so on. Eventually, all the gases in air can be made to liquefy (change into a liquid).

But the reverse process also takes place. Suppose a container of liquid air holds 100 liters. The liquid air will warm up slowly. When its temperature reaches −195.79°C, liquid nitrogen changes back to a gas. Since about 78 percent of air is nitrogen, only 22 percent of the original liquid air (22 liters) will be left.

When the temperature reaches −182.96°C, oxygen changes from a liquid back to a gas. Since oxygen makes up 21 percent of air, another 21 percent (21 liters) of the liquid air will evaporate.

The work of Ramsay and Travers was very difficult, however, because the gases they were looking for are not abundant in air. Krypton, for example, makes up only about 0.000114 percent of air. For every 100 liters of liquid air, there would be only 0.00011, or about one-tenth of a milliliter of krypton. A tenth of a milliliter is about a drop. So Ramsay and Travers—

WORDS TO KNOW

Isotopes two or more forms of an element that differ from each other according to their mass number

Noble gas an element in Group 18 (VIIIA) of the periodic table

Periodic table a chart that shows how the chemical elements are related to each other

Phosphor material that gives off light when struck by electrons

Radioactive isotope an isotope that breaks apart and gives off some form of radiation

"Look, up in the sky! It's a bird! It's a plane....

The famous cartoon character Superman has many super powers. Everybody knows that. He's the Man of Steel. He has X-ray vision. His hearing is so good, he can tune in on one voice in a crowded city. And, of course: He's faster than a speeding bullet! More powerful than a locomotive! Able to leap tall buildings in a single bound!

But there's one substance that weakens Superman: kryptonite! If exposed to kryptonite, Superman experiences pain and loses his super powers. If exposed for too long, he can even die.

Kryptonite, of course, is purely fictional. Despite the similarity in names, kryptonite has nothing to do with element 36, krypton. According to cartoon legend, Superman came from the planet Krypton. Kal-El, as he was originally known, was placed in a spaceship by his parents, moments before the planet exploded.

Unfortunately, as the young Superman blasted away from Krypton, a piece of kryptonite got stuck on the spaceship. The same terrible forces that caused the planet to explode, also had created the deadly kryptonite. And, as Superman would later find out, arch-villains always seem to get their hands on this green glowing rock!

Aside from the fictitious nature of kryptonite, there is another difference between it and krypton. Kryptonite is a rock—one that can cause great harm to, well, one person anyway. Krypton is an inert gas that has no effect on anything.

although they didn't know it—were looking for one drop of krypton in 100 liters of liquid air!

Amazingly, they found it. The discovery of these three gases was a great credit to their skills as researchers. They suggested the name krypton for the new element. The name was taken from the Greek word *kryptos* for "hidden."

Physical properties

Krypton is a colorless, odorless gas. It has a boiling point of –152.9°C (–243.2°F) and a density of 3.64 grams per liter. That makes krypton about 2.8 times as dense as air.

Chemical properties

For many years, krypton was thought to be completely inert. Then, in the early 1960s, it was found to be possible to make certain compounds of the element. English chemist Neil Bartlett (1932–) found ways to combine noble gases with the most active element of all, **fluorine.** In 1963, the first krypton compounds were made—krypton difluoride (KrF_2) and krypton tetrafluoride (KrF_4). Other compounds of krypton have also been made since that time. However, these have no commercial uses. They are only laboratory curiosities.

Occurrence in nature

The abundance of krypton in the atmosphere is thought to be about 0.000108 to 0.000114 percent. The element is also formed in the Earth's crust when **uranium** and other radioactive elements break down. The amount in the Earth's crust is too small to estimate, however.

Isotopes

Six naturally occurring isotopes of krypton exist. They are krypton-78, krypton-80, krypton-82, krypton-83, krypton-84, and krypton-86. Isotopes are two or more forms of an element. Isotopes differ from each other according to their mass number. The number written to the right of the element's name is the mass number. The mass number represents the number of protons plus neutrons in the nucleus of an atom of the element. The number of protons determines the element, but the number of neutrons in the atom of any one element can vary. Each variation is an isotope.

At least sixteen radioactive isotopes of krypton are known also. A radioactive isotope is one that breaks apart and gives off some form of radiation. Radioactive isotopes are produced when very small particles are fired at atoms. These particles stick in the atoms and make them radioactive.

One radioactive isotope of krypton is used commercially, krypton-85. It can be combined with phosphors to produce materials that shine in the dark. A phosphor is a material that shines when struck by electrons. Radiation given off by krypton-85 strikes the phosphor. The phosphor then gives off light. The same isotope is also used for detecting leaks in a container. The radioactive gas is placed inside the container to be tested. Since the gas is inert, krypton will not react with anything else in the container. But if the container has a leak, some radioactive krypton-85 will escape. The isotope can be detected with special devices for detecting radiation.

Krypton-85 is also used to study the flow of blood in the human body. It is inhaled as a gas, and then absorbed by the blood. It travels through the bloodstream and the heart along

How long is a meter?

The meter is the standard unit of length in the metric system. It was first defined in 1791. As part of the great changes brought by the French Revolution, an entirely new system of measurement was created: the metric system.

At first, the meter was defined in a very simple way. It was the distance between two lines scratched into a metal bar kept outside Paris. For many years, that definition was satisfactory for most purposes. Of course, it created a problem. Suppose someone in the United States was in the business of making meter sticks. That person would have to travel to Paris to make a copy of the official meter. Then the copy would have to be used to make other copies. The chances for error in this process are tremendous.

In 1960, scientists had another idea. They suggested using light produced by hot krypton as the standard of length. Here is how that standard was developed:

When an element is heated, it absorbs energy from the heat. The atoms present in the element are in an "excited," or energetic, state. Atoms normally do not remain in an excited state very long. They give off the energy they just absorbed and return to their normal, "unexcited" state. The energy they give off can take different forms. One of those forms is light.

The kind of light given off is different for each element and for each isotope. The light usually consists of a series of very bright lines called a spectrum. The number and color of the lines produced is specific to each element and isotope.

When one isotope of krypton, krypton-86, is heated, it gives off a very clear, distinct, bright line with a reddish-orange color. Scientists decided to define the meter in terms of that line. They said that a meter is 1,650,763.73 times the width of that line.

This standard had many advantages. For one thing, almost anyone anywhere could find the official length of a meter. All one needed was the equipment to heat a sample of krypton-86. Then one had to look for the reddish-orange line produced. The length of the meter, then, was 1,650,763.73 times the width of that line.

This definition for the meter lasted only until 1983. Scientists then decided to define a meter by how fast light travels in a vacuum. This system is even more exact than the one based on krypton-86.

with the blood. Its pathway can be followed by a technician who holds a detection device over the patient's body. The device shows where the radioactive material is going and how fast it is moving. A doctor can determine whether this behavior is normal or not.

Extraction

Krypton is still obtained by allowing liquid air to evaporate.

Uses

The only commercial uses of krypton are in various kinds of lamps. When an electric current is passed through krypton gas, it gives off a very bright light. Perhaps the most common application of this principle is in airport runway lights. These lights are so bright that they can be seen even in foggy conditions for distances up to 300 meters (1,000 feet). The lights do not burn continuously. Instead, they send out very brief pulses of light. The pulses last no more than about 10 microseconds (10 millionths of a second). They flash on and off about 40 times per minute. Krypton is also used in slide and movie projectors.

Although neon lights sometimes *do* include neon, krypton is often the gas used.

Krypton gas is also used in making "neon" lights. Neon lights are colored lights often used in advertising. They are similar to fluorescent light bulbs. But they give off a colored light because of the gas they contain. Some neon lights *do* contain the gas neon, but others contain other noble gases. A neon light filled with krypton, for example, glows yellow.

Compounds

Compounds of krypton have been prepared in the laboratory but do not exist in nature. The synthetic (artificial) compounds are used for research purposes only.

Health effects

There is no evidence that krypton is harmful to humans, animals, or plants.

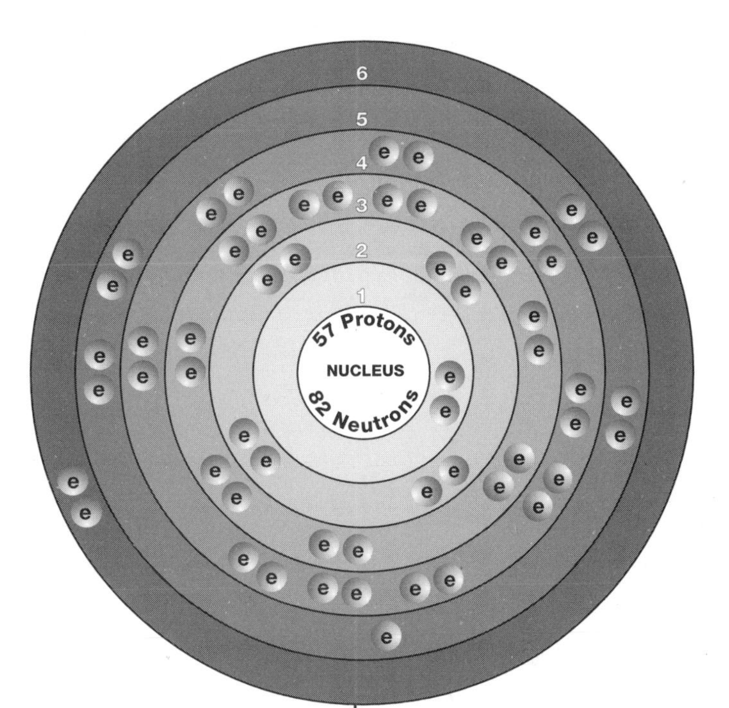

LANTHANUM

Overview

Lanthanum is the third element in Row 6 of the periodic table. The periodic table is a chart that shows how chemical elements are related to each other. Lanthanum is a transition element in Group 3 (IIIB) of the periodic table. Lanthanum's position makes it one of the transition metals. The transition metals are found in the center of the periodic table, in Groups 3 through 12.

Lanthanum can also be classified as a rare earth element. The rare earth elements are the 15 elements that make up Row 6 of the periodic table. That group of elements is also called the lanthanides. Either way of classifying lanthanum is acceptable to most chemists.

Lanthanum was first discovered by Swedish chemist Carl Gustav Mosander (1797–1858) in 1839. This discovery was the first chapter in a long and interesting story. At the end of that story, six more elements had been discovered. All of these elements occur together in nature and are hard to separate from each other.

SYMBOL
La

ATOMIC NUMBER
57

ATOMIC MASS
138.9055

FAMILY
Group 3 (IIIB)
Transition metal

PRONUNCIATION
LAN-tha-num

WORDS TO KNOW

Alloy a mixture of two or more metals with properties different from those of the individual metals

Carbon arc lamp a lamp for producing very bright white light

Ductile capable of being drawn into thin wires

Half life the time it takes for half of a sample of a radioactive element to break down

Isotopes two or more forms of an element that differ from each other according to their mass number

Lanthanides the elements that make up Row 6 of the periodic table between barium and hafnium

Malleable capable of being hammered into thin sheets

Periodic table a chart that shows how the chemical elements are related to each other

Phosphor a material that gives off light when struck by electrons

Radioactivity having a tendency to break down and give off radiation

Rare earth elements *see* **Lanthanides**

Lanthanum metal has relatively few uses. Some of its compounds however, are used in lamps, color television sets, cigarette lighters, and optical fibers.

Discovery and naming

Toward the end of the 1830s, Mosander became interested in an unusual black stone found near the town of Bastnas, Sweden. He found that the stone contained two new materials. He thought those materials were both new elements. Mosander called them **cerium** and lanthanum. He was right about cerium, but wrong about lanthanum. The material Mosander called lanthanum later turned out to be a mixture of six new elements.

It took scientists more than 60 years to sort out these elements and separate them from each other. It was not until 1923 that a pure sample of lanthanum metal was even prepared. Still, Mosander is given credit for the discovery of lanthanum.

Physical properties

Lanthanum is a white, ductile, malleable metal. Ductile means capable of being drawn into thin wires. Malleable means capable of being hammered into thin sheets. It is relatively soft and can be cut with a sharp knife. Its melting point is 920°C (1,690°F) and its boiling point is 3,454°C (6,249°F). Its density is about 6.18 grams per cubic centimeter.

Chemical properties

Lanthanum is a very active metal. It reacts with most acids and with cold water, although slowly. With hot water, it reacts more quickly:

$$2La + 6H_2O \rightarrow 2La(OH)_3 + 3H_2$$

Lanthanum also reacts with **oxygen** in the air, especially if the air is moist.

Occurrence in nature

Lanthanum is relatively common in the Earth's crust. Its abundance is thought to be as high as 18 parts per million. That would make it nearly as common as **copper** or **zinc.** Unlike those metals, however, it usually does not occur in one place, as in copper mines. Instead, it is spread widely throughout the Earth's crust. Its most common minerals are monazite, bastna-

Magnified view of lanthanum aluminate crystal.

site, and cerite. These minerals generally contain all the other rare earth elements as well.

Isotopes

Two naturally occurring isotopes of lanthanum are known. They are lanthanum-138 and lanthanum-139. Isotopes are two or more forms of an element. Isotopes differ from each other according to their mass number. The number written to the right of the element's name is the mass number. The mass number represents the number of protons plus neutrons in the nucleus of an atom of the element. The number of protons determines the element, but the number of neutrons in the atom of any one element can vary. Each variation is an isotope.

Lanthanum-138 is very rare and is radioactive. Its half life is about 100 billion years. The half life of a radioactive element is the time it takes for half of a sample of the element to break down. Only 5 grams of a 10-gram sample of lanthanum-138

will remain after 100 billion years. The other 5 grams would have broken down to form a new isotope.

More than a dozen artificial radioactive isotopes have also been made. These isotopes are produced when very small particles are fired at atoms. These particles stick in the atoms and make them radioactive. None of the radioactive isotopes of lanthanum have any commercial use.

Extraction

The rare earth elements are very similar to each other. Separating them is a very difficult task. The ores are first treated with sulfuric acid (H_2SO_4). The materials produced are then passed through a series of steps and the individual elements separated from each other.

Lanthanum oxide is used to make phosphors. Phosphors give off light when struck by electrons. They produce the colors on television sets.

Uses and compounds

One of the most important uses of lanthanum compounds is in carbon arc lamps. In a carbon arc lamp, an electrical current is passed through the lamp electrode. The electrode is made of carbon and traces of other materials that have been added. The current causes the carbon to heat up and give off a brilliant white light. The exact color of the light depends on the other materials that have been added to the carbon. Lanthanum fluoride (LaF_3) and lanthanum oxide (La_2O_3) are usually used for this purpose.

These two compounds are also used in making phosphors. A phosphor is a material that gives off light when struck by electrons. The color of the phosphor depends on what elements are present. The colors produced in a color television set are caused by phosphors painted on the back of the screen.

Compounds of lanthanum are also used to make special kinds of glass. High quality lenses, for example, are often made of glass containing a small amount of lanthanum.

One of the oldest uses of lanthanum metals is in the production of misch metal. Misch metal is an alloy that produces sparks when struck. One application of misch metal is in the manufacture of cigarette lighters.

A new application of lanthanum glass is in making optical fibers. An optical fiber is a wire-like material made of glass. It

carries light in the same way a copper wire carries electricity. Optical fibers are becoming more popular as methods for carrying audio, video, and digital messages. In many cases, optical fibers have replaced copper wires for this purpose.

In 1974, a team of French researchers invented a product called ZBLAN. ZBLAN is made of **zirconium, barium,** lanthanum, **aluminum,** and **sodium.** ("ZBLAN" comes from the first letters of each element's symbol.) Scientists have found that ZBLAN is 100 times better at carrying messages than traditional optical fibers.

Health effects

Lanthanum and its compounds are poisonous in high concentrations. They should be handled with care.

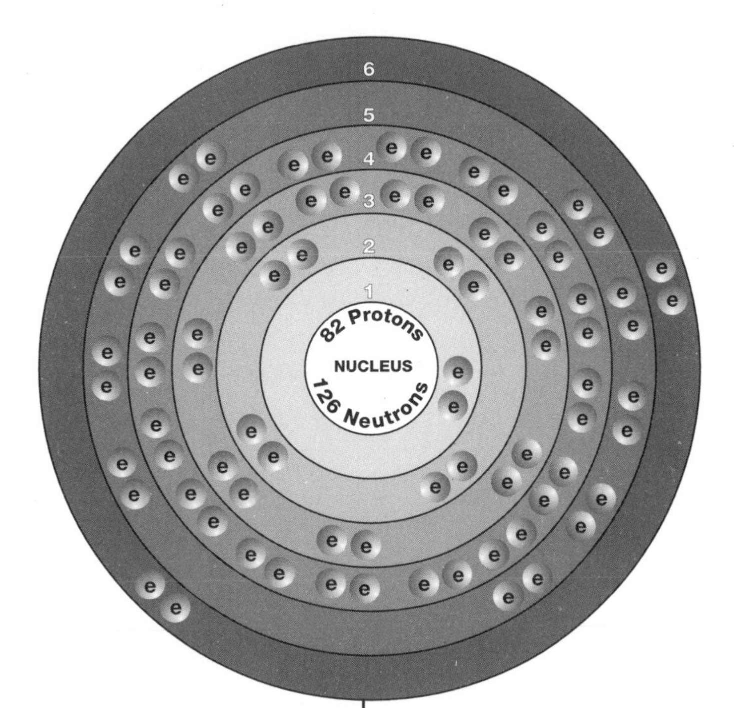

LEAD

Overview

Lead is the heaviest member of the **carbon** family. The carbon family consists of the five elements in Group 14 (IVA) of the periodic table. The periodic table is a chart that shows how chemical elements are related to each other. Although a member of the carbon family, lead looks and behaves very differently from carbon.

Lead is one of only a few elements known to ancient peoples. One of the oldest examples of lead is a small statue found in Egypt. It was made during the First Dynasty, in about 3400 B.C. Mention of lead and lead objects can also be found in very old writing from India. And the Bible mentions lead in a number of passages.

Throughout history, lead has been used to make water and sewer pipes; roofing; cable coverings; type metal and other alloys; paints; wrappings for food, tobacco, and other products; and as an additive in gasoline. Since the 1960s, however, there has been a growing concern about the health effects of lead. For instance, scientists have found that lead can cause mental and physical problems in growing children. As a result, many common lead products are now being phased out.

SYMBOL
Pb

ATOMIC NUMBER
82

ATOMIC MASS
207.2

FAMILY
Group 14 (IVA)
Carbon

PRONUNCIATION
LED

Discovery and naming

Lead has been around for thousands of years. It is impossible to say when humans first discovered the element. It does not occur as an element in the earth very often. But one of its ores, lead sulfide (PbS), is fairly common. It is not difficult to obtain pure lead metal from lead sulfide. Humans probably discovered methods for doing so thousands of years ago.

By Roman times, lead metal was widely used. The far-reaching system that brought water to Rome contained many lead pipes. Sheets of lead were used as writing tablets and some Roman coins were also made of lead. Perhaps of greatest interest was the use of lead in making pots and pans. Modern scientists believe many Romans may have become ill and died because of this practice. Cooking liquids in lead utensils tends to make the lead dissolve. It got into the food being cooked. People who ate those foods got more and more lead into their bodies. Eventually, the effects of lead poisoning must have begun to appear.

Of course, the Romans had little understanding of the connection between lead and disease. They probably never realized that they were poisoning themselves by using lead pots and pans.

No one is quite sure how lead got its name. The word has been traced to manuscripts that date to before the 12th century. Romans called the metal *plumbum*. It is from this name that the element's chemical symbol comes: Pb. Compounds of lead are sometimes called by this old name, such as plumbous chloride.

Physical properties

Lead is a heavy, soft, gray solid. It is both ductile and malleable. Ductile means capable of being drawn into thin wires. Malleable means capable of being hammered into thin sheets. It has a shiny surface when first cut, but it slowly tarnishes (rusts) and becomes dull. Lead is easily worked. "Working" a metal means bending, cutting, shaping, pulling, and otherwise changing the shape of the metal.

The melting point of lead is 327.4°C (621.3°F), and its boiling point is 1,750 to 1,755°C (3,180 to 3,190°F). Its density is 11.34 grams per cubic centimeter. Lead does not conduct an electric current, sound, or vibrations very well.

WORDS TO KNOW

Ductile capable of being drawn into thin wires

Isotopes two or more forms of an element that differ from each other according to their mass number

Laser a device for making very intense light of one very specific color that is intensified many times over

Malleable capable of being hammered into thin sheets

Periodic table a chart that shows how the chemical elements are related to each other

Radioactive isotope an isotope that breaks apart and gives off some form of radiation

Toxic poisonous

Chemical properties

Lead is a moderately active metal. It dissolves slowly in water and in most cold acids. It reacts more rapidly with hot acids. It does not react with **oxygen** in the air readily and does not burn.

Occurrence in nature

The abundance of lead in the Earth's crust is estimated to be between 13 and 20 parts per million. It ranks in the upper third among the elements in terms of its abundance.

Lead rarely occurs as a pure element in the earth. Its most common ore is galena, or lead sulfide (PbS). Other ores of lead are anglesite, or lead sulfate ($PbSO_4$); cerussite, or lead carbonate ($PbCO_3$); and mimetite ($PbCl_2 \bullet Pb_3(AsO_4)_2$).

The largest producers of lead ore in the world are Australia, China, the United States, Peru, Canada, Mexico, and Sweden. In the United States, more than 93 percent of all the lead produced comes from Missouri. Other lead-producing states are Montana, Colorado, Idaho, Illinois, New York, and Tennessee. In 1996, 426,000 metric tons of lead were produced in the United States.

Romans routinely ate food cooked in lead pots and pans. The connection between lead and disease was not known then, so many people became ill and died of lead poisoning.

Isotopes

Four naturally occurring isotopes of lead occur. They are lead-204, lead-206, lead-207, and lead-208. Isotopes are two or more forms of an element. Isotopes differ from each other according to their mass number. The number written to the right of the element's name is the mass number. The mass number represents the number of protons plus neutrons in the nucleus of an atom of the element. The number of protons determines the element, but the number of neutrons in the atom of any one element can vary. Each variation is an isotope.

About sixteen radioactive isotopes of lead are known also. A radioactive isotope is one that breaks apart and gives off some form of radiation. Radioactive isotopes are produced when very small particles are fired at atoms. These particles stick in the atoms and make them radioactive.

One radioactive isotope of lead, lead-210, is sometimes used in medicine. This isotope gives off radiation that can kill cancer cells. It is also used to treat non-cancerous eye disorders.

Lead smelting.

Extraction

Lead is obtained from its ores by a method used with many metals. First, the ore is roasted (heated in air). Roasting, also called smelting, converts the ore to a compound of lead and oxygen, lead oxide (PbO_2). Lead oxide is then heated with charcoal (pure carbon). The carbon takes oxygen away from the lead oxide. It leaves pure lead behind:

$$PbO_2 + C \xrightarrow{\text{heated}} CO_2 + Pb$$

A major source of lead is recycled car batteries.

Lead obtained in this way is not very pure. It can be purified electrolytically. Electrolytic refining involves passing an electric current through a compound. Very pure lead is collected at one side of the container in which the reaction is carried out.

Lead is also recovered in recycling programs. Recycling is the process by which a material is retrieved from a product that is no longer used. For example, old car batteries were once just thrown away. Now they are sent to recycling plants where lead

The price of a gallon of gas

For many years, lead was regarded as a miracle chemical by the automotive industry. The power to run a car comes from the burning of gasoline in the engine. However, burning gasoline is not a simple process. Many things happen inside an engine when gasoline burns in the carburetor.

For example, an engine can "knock" if the gasoline does not burn properly. "Knocking" is a "bang-bang" sound from the engine. It occurs when low-grade gasoline is used.

One way to prevent knocking is to use high-grade gasoline. Another way is to add chemicals to the gasoline. The best gasoline additive discovered was a compound called tetraethyl lead ($Pb(C_2H_5)_4$). Tetraethyl lead was usually called "lead" by the automotive industry, the consumer, and advertisers. When someone bought "leaded" gasoline, it contained not lead metal, but tetraethyl lead.

Leaded gasoline was a great discovery. It could be made fairly cheaply and it prevented car engines from knocking. No wonder people thought it was a miracle chemical.

What people didn't realize was that tetraethyl lead breaks down in a car engine because of the high temperature at which engines operate. When tetraethyl lead breaks down, elemental lead (Pb) is formed:

$$Pb(C_2H_5)_4 \xrightarrow{\text{heat}} Pb + \text{other products}$$

The result—with millions of cars being driven every day—was more and more lead getting into the air. And more and more people inhaled that lead. Eventually, doctors began to see more people with lead-related diseases.

The federal government finally decided that tetraethyl lead was too dangerous to use in gasoline. By 1990, the use of this compound had been banned by all governments in North America.

can be extracted and used over and over again. It is not necessary to get all the lead that industry needs from new sources, such as ores.

Uses

The lead industry is undergoing dramatic change. Many products once made with lead no longer use the element. The purpose of this change is to reduce the amount of lead that gets into the environment. Examples of such products include ammunition, such as shot and bullets; sheet lead used in building construction; solder; water and sewer pipes; ball bearings; radiation shielding; and gasoline. These changes are possible because manufacturers are finding safer elements to use in place of lead.

Other uses of lead have not declined. The best example is lead storage batteries. A lead storage battery is a device for converting chemical energy into electrical energy. Almost every car and truck has at least one lead storage battery. But no satisfactory substitute for it has been found. About 87 percent of all lead produced in the United States now goes to the manufacture of lead storage batteries. In addition to cars and trucks, these batteries are used for communication networks and emergency power supplies in hospitals, and in forklifts, airline ground equipment, and mining vehicles.

Compounds

A small percentage of lead is used to make lead compounds. Although the amount of lead is small, the variety of uses for these compounds is large. Some examples of important lead compounds are:

lead acetate ($Pb(C_2H_3O_2)_2$): insecticides; waterproofing; varnishes; dyeing of cloth; production of gold; hair dye

lead antimonate ($Pb_3(SbO_4)_2$): staining of glass, porcelain and other ceramics

lead azide ($Pb(N_3)_2$): used as a "primer" for high explosives

lead chromate ("chrome yellow"; $PbCrO_4$): industrial paints (use restricted by law)

lead fluoride (PbF_2): used to make lasers; specialized optical glasses

lead iodide (PbI_2): photography; cloud seeding to produce rain

lead naphthenate ($Pb(C_7H_{12}O_2)$): wood preservative; insecticide; additive for lubricating oil; paint and varnish drier

lead phosphite ($2PbO \bullet PbHPO_3$): used to screen out ultraviolet radiation in plastics and paints

lead stearate ($Pb(C_{18}H_{35}O_2)_2$): used to make soaps, greases, waxes, and paints; lubricant; drier for paints and varnishes

lead telluride (PbTe): used to make semiconductors, photoconductors, and other electronic equipment

Health effects

The health effects of lead have become much better understood since the middle of the 20th century. At one time, the metal was regarded as quite safe to use for most applications. Now lead is known to cause both immediate and long-term health problems, especially with children. It is toxic when swallowed, eaten, or inhaled.

Young children are most at risk from lead poisoning. Some children have a condition known as pica. They have an abnormal desire to eat materials like dirt, paper, and chalk. Children with pica sometimes eat paint chips off walls. At one time, many interior house paints were made with lead compounds. Thus, crawling babies or children with pica ran the risk of eating large amounts of lead and being poisoned.

Lead causes both immediate and long-term health problems, especially with children. It is toxic when swallowed, eaten, or inhaled.

Some symptoms of lead poisoning include nausea, vomiting, extreme tiredness, high blood pressure, and convulsions (spasms). Over a long period of time, these children often suffer brain damage. They lose the ability to carry out normal mental functions.

Other forms of lead poisoning can also occur. For example, people who work in factories where lead is used can inhale lead fumes. The amount of fumes inhaled at any one time may be small. But over months or years, the lead in a person's body can build up. This kind of lead poisoning can lead to nerve damage and problems with the gastrointestinal system (stomach and intestines).

Today, there is an effort to reduce the use of lead in consumer products. For instance, older homes are often tested for lead paint before they are resold. Lead paint has also been removed from older school buildings.

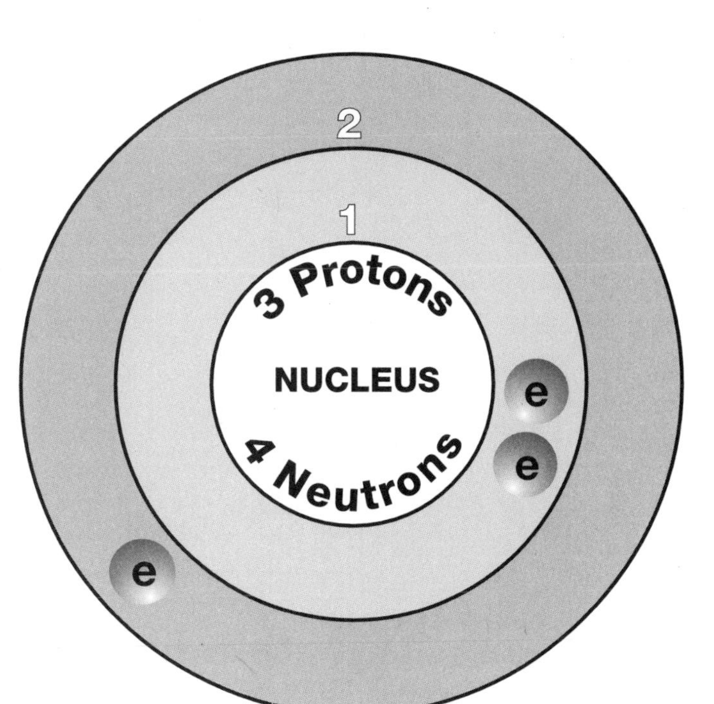

LITHIUM

Overview

Lithium is the first member of the alkali metal family. The alkali metals are the elements that make up Group 1 (IA) of the periodic table. The periodic table is a chart that shows how chemical elements are related to one another. The alkali metals include **sodium, potassium, rubidium, cesium,** and **francium.** Lithium is also the least dense of all metals. It has a density about half that of water.

Credit for the discovery of lithium usually goes to Swedish chemist Johan August Arfwedson (or Arfvedson; 1792–1841). Arfwedson found the new element in a mineral that had first been identified about twenty years earlier by Brazilian scientist Jozé Bonifácio de Andrada e Silva (1763–1838). That mineral, petalite, is still a major source of lithium today.

Lithium has a number of important and interesting uses. In recent years, it has been used to make lightweight, efficient batteries. Compounds of lithium have also been used to treat a mental disorder known as bipolar disorder.

Discovery and naming

The first clues to the existence of lithium surfaced in 1800. De Andrada was a Brazilian scientist and statesman visiting in

SYMBOL
Li

ATOMIC NUMBER
3

ATOMIC MASS
6.941

FAMILY
Group 1 (IA)
Alkali metal

PRONUNCIATION
LI-thee-um

Scandinavia. During one of his trips to the countryside, he came across a mineral that he did not recognize. He called the mineral petalite.

Some scientists were not convinced that petalite was a new mineral. But in 1817, the same mineral was rediscovered on the island of Utö. Interest in the mineral grew.

Arfwedson was troubled by the results of his analysis of petalite. In his studies, he could not identify 10 percent of the mineral. He finally concluded that the missing 10 percent must be a new element. He called the new element lithium, from the Greek word *lithos* for "stone."

Arfwedson was not able to produce pure lithium. About a year later, however, Swedish chemist William Thomas Brande (1788–1866) and English chemist Sir Humphry Davy (1778–1829) were both able to extract the pure metal from its compounds. (See sidebar on Davy in the **calcium** entry in Volume 1.)

Physical properties

Lithium is a very soft, silvery metal. It has a melting point of 180.54°C (356.97°F) and a boiling point of about 1,335°C (2,435°F). Its density is 0.534 grams per cubic centimeter. By comparison, the density of water is 1.000 grams per cubic centimeter. Lithium's hardness on the Mohs scale is 0.6. The Mohs scale is a way of expressing the hardness of a material. It runs from 0 (for talc) to 10 (for diamond). A hardness of 0.6 means that the material can be scratched with a fingernail.

Chemical properties

Lithium is an active element, but not as active as the other alkali metals. It reacts slowly with water at room temperature and more rapidly at higher temperatures. It also reacts with most acids, giving off **hydrogen** gas. Lithium does not react with **oxygen** at room temperature, but above 100°C does so to form lithium oxide (Li_2O). Under the proper conditions, the element also combines with sulfur, hydrogen, nitrogen, and the halogens.

Occurrence in nature

The abundance of lithium in the Earth's crust is estimated to be about 0.005 percent. That places it among the top 15 ele-

WORDS TO KNOW

Alkali metals elements that make up Group 1 (IA) of the periodic table

Alloy a mixture of two or more metals with properties different from those of the individual metals

Bipolar disorder a condition in which a person experiences wild mood swings

Catalyst a substance used to speed up or slow down a chemical reaction without undergoing any change itself

Isotopes two or more forms of an element that differ from each other according to their mass number

Periodic table a chart that shows how the chemical elements are related to each other

Radioactive isotope an isotope that breaks apart and gives off some form of radiation

Lithium metal.

ments found in the earth. The most common ores of lithium are spodumene, petalite, and lepidolite. Lithium is also obtained from saltwater. As saltwater evaporates, dissolved solids are left behind. These solids include sodium chloride (NaCl), potassium chloride (KCl), and lithium chloride (LiCl).

The world's largest producer of lithium is the United States. Three of the largest U.S. mines are located in Silver Peak, Nevada, and Kings Mountain and Bessemer City, North Carolina. Other major producers of lithium compounds are Australia, Russia, Canada, Zimbabwe, Chile, and China.

Isotopes

Two naturally occurring isotopes of lithium exist, lithium-6 and lithium-7. Isotopes are two or more forms of an element. Isotopes differ from each other according to their mass number. The number written to the right of the element's name is the mass number. The mass number represents the number of protons plus neutrons in the nucleus of an atom of the element. The number of protons determines the element, but the number of neutrons in the atom of any one element can vary. Each variation is an isotope.

In addition, three radioactive isotopes of lithium have been produced. A radioactive isotope is one that breaks apart and gives off some form of radiation. Radioactive isotopes are produced when very small particles are fired at atoms. These particles stick in the atoms and make them radioactive. None of these isotopes has any important commercial application.

Lithium carbonate is added to glass to make it stronger. Pyrex cookware is made up of this kind of glass.

Extraction

Lithium compounds are first converted to lithium chloride (LiCl). Then, an electric current is passed through molted (melted) lithium chloride. The current separates the compound into lithium and **chlorine** gas:

$$2LiCl \xrightarrow{\text{electric current}} 2Li + Cl_2$$

Uses and compounds

Lithium metal and its compounds have a great many uses. Two of the most significant applications are in the glass and ceramics field and in the production of **aluminum.** The addition of a small amount of lithium carbonate (Li_2CO_3) to a glass or ceramic makes the material stronger. Examples of the use of lithium carbonate are shock-resistant cookware (such as the Pyrex brand) and black-and-white television tubes. About 40 percent of the lithium used in the United States in 1996 went to these applications.

Producers of aluminum also use lithium carbonate in preparing aluminum metal from aluminum oxide. Lithium carbonate reduces the heat needed to make the reaction occur. As a result, producers save money by using less energy. In 1996, about 20 percent of all lithium carbonate produced in the United States went to this application.

Feeling better with lithium

An exciting new use for lithium carbonate was discovered in 1949. John Cade (1912–80), an Australian physician, found that patients with bipolar disorder benefitted from taking lithium carbonate. Bipolar disorder is a condition once known as manic-depressive disorder. The condition is characterized by dramatic mood swings. A person can be very happy and carefree one moment, but terribly depressed the next moment. Some patients become so depressed that they commit suicide. Until 1949, there was no effective treatment for bipolar disorder.

Cade found that most patients who took lithium carbonate were relieved of at least some of their symptoms. Their "high" points were not as high, and their "low" points were not as low. The compound helped someone with bipolar disorder to live a quieter, more normal life. Today, more than 60 percent of those with bipolar disorder benefit from lithium treatments.

As with most medications, lithium compounds can have side effects. They can cause nausea, dizziness, diarrhea, dry mouth, and weight gain. But these side effects can usually be controlled. And they are often a small price to pay for relief from the terrible effects of bipolar disorder.

Another important compound of lithium is lithium stearate. Lithium stearate is added to petroleum to make a thick lubricating grease. The grease is used in many industrial applications because it does not break down at high temperatures, it does not become hard when cooled, and it does not react with water or oxygen in the air. Lithium greases are used in military, industrial, automotive, aircraft, and marine applications. Lithium stearate is also used as an additive in cosmetics and plastics. Overall, the manufacture of lithium stearate is the third most important use of lithium compounds after glasses and ceramics manufacture and aluminum production.

The first commercial use of lithium was in the production of alloys. An alloy is made by melting and mixing two or more metals. The mixture has properties different from those of the individual metals. Early lithium alloys included lead and were used to make tough ball bearings for machinery.

Today, the most commonly used alloys of lithium are made with aluminum or **magnesium.** These alloys are very light, but very strong. They are used for armor plates and in aerospace applications.

Lithium compounds are also used as catalysts in many different industrial processes. A catalyst is a substance used to speed up or slow down a chemical reaction. The catalyst does not undergo any change itself during the reaction. For example, one lithium catalyst is used to make tough, strong, synthetic (artificial) rubber. It does not have to be vulcanized (heat-treated) like natural rubber.

Lithium has become important in the manufacture of batteries. A battery is a device for converting chemical energy into electrical energy. Car batteries use a chemical reaction between lead and sulfuric acid to make electrical energy.

Lithium compounds tend to harm the kidneys.

Lithium batteries are much lighter than lead and sulfuric acid batteries. They also reduce the use of toxic lead and cadmium. Lithium batteries are used in products such as watches, microcomputers, cameras, small appliances, electronic games, toys, and many kinds of military and space vehicles.

Health effects

Lithium and its compounds have a range of effects on the human body. For instance, compounds of lithium tend to harm the kidneys. And lithium carbonate (Li_2CO_3) can affect a person's mental health (see accompanying sidebar).

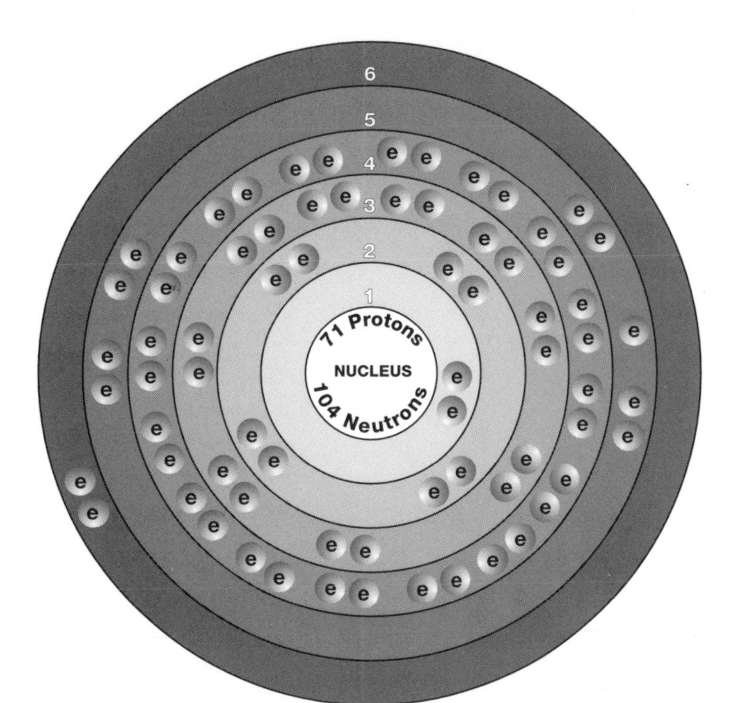

LUTETIUM

Overview

Lutetium is the heaviest, rarest, and most expensive lanthanide element. The lanthanide elements make up Row 6 of the periodic table. The periodic table is a chart that shows how chemical elements are related to one another. The lanthanides are pulled out into a separate row at the bottom of the table. They are also called the rare earth elements. That name does not fit very well for most lanthanides. They are not really so rare, only difficult to separate from each other. However, lutetium is both rare and difficult to separate from the other lanthanides.

Lutetium was first discovered in the early 1900s by two chemists working independently. It was found in a complex black mineral that had been found near the town of Ytterby, Sweden, in 1787.

Today, there are very few uses for lutetium metal.

SYMBOL
Lu

ATOMIC NUMBER
71

ATOMIC MASS
174.97

FAMILY
Lanthanide
(rare earth metal)

PRONUNCIATION
loo-TEE-she-um

Discovery and naming

In 1787, a Swedish army officer, Carl Axel Arrhenius (1757–1824), found an odd black rock outside the town of Ytterby, Sweden. He gave the rock to a chemist friend, Johan

Gadolin (1760–1852), for study. That rock turned out to contain one of the most complex and most interesting minerals ever discovered: yttria. Chemists kept busy for the next century trying to figure out exactly what yttria was made of.

Eventually, they found nine new elements that had never been seen before. Separating these elements from each other was very difficult, however. The nine elements are chemically similar and all behave in nearly the same way. It is very difficult to know whether a sample of yttria contains one, two, three . . . or all nine of the elements.

In 1879 French chemist Jean-Charles-Galissard de Marignac (1817–94) announced the discovery of a new element in yttria. He called the element **ytterbium.** Other chemists suspected that ytterbium was really a mixture of elements. They searched for ways to separate ytterbium into simpler parts.

It took nearly thirty years to solve this puzzle. And the answer came from three laboratories at nearly the same time. The first to report his results was French chemist Georges Urbain (1872–1938). In 1907, he reported that ytterbium was not an element, but a mixture of two new elements. He called those elements neoytterbium and lutecium. The first name meant "new ytterbium." The second name comes from Lutecia, the ancient name for the city of Paris.

At nearly the same time, German chemist Karl Auer (Baron von Welsbach; 1858–1929) made the same discovery. He suggested different names for the two new elements in ytterbium. He called them cassiopeium and aldebaranium, in honor of the constellation Cassiopeia and the bright star Aldebaran. Today, some German chemists still refer to lutetium as cassiopeium.

A third chemist working on ytterbium was American chemist Charles James (1880–1926). James announced his discoveries after Urbain and Auer. Some authorities give credit for the discovery of lutetium to all three scientists.

None of these early scientists actually saw pure lutetium. Their element was a compound, usually lutetium oxide. The pure metal was isolated only quite recently.

In 1949, the spelling of the element changed from "lutecium" to "lutetium."

WORDS TO KNOW

Catalyst a substance used to speed up or slow down a chemical reaction without undergoing any change itself

Ductile capable of being drawn into a thin wire

Isotopes two or more forms of an element that differ from each other according to their mass number

Lanthanides the elements in the periodic table with atomic numbers 58 through 71

Radioactive isotope an isotope that breaks apart and gives off some form of radiation

Rare earth element *see* **Lanthanides**

Toxic poisonous

Physical properties

Lutetium is a silvery white metal that is quite soft and ductile. The term ductile means capable of being drawn into thin wires. It has a melting point of 1,652°C (3,006°F) and a boiling point of 3,327°C (6,021°F). Its density is 8.49 grams per cubic centimeter.

Chemical properties

Lutetium reacts slowly with water and dissolves in acids. Other chemical properties tend to be of interest only to researchers.

Occurrence in nature

Lutetium is thought to be very rare in the Earth's crust. It occurs to the extent of about 0.8 to 1.7 parts per million. That still makes it somewhat more common than better known elements such as **iodine, silver,** and **mercury.** The most common ore of lutetium is monazite, in which its concentration is about 0.003 percent.

Although very rare, lutetium is still more common than iodine, silver, and mercury.

Isotopes

There are two naturally occurring isotopes of lutetium, lutetium-175 and lutetium-176. Isotopes are two or more forms of an element. Isotopes differ from each other according to their mass number. The number written to the right of the element's name is the mass number. The mass number represents the number of protons plus neutrons in the nucleus of an atom of the element. The number of protons determines the element, but the number of neutrons in the atom of any one element can vary. Each variation is an isotope.

The second of these isotopes, lutetium-176, is radioactive. A radioactive isotope is one that breaks apart and gives off some form of radiation. Some radioactive isotopes occur in nature. Others can be produced by firing very small particles at atoms. These particles stick in the atoms and make them radioactive.

Fourteen artificial radioactive isotopes have also been produced. They have atomic masses of 155, 156, 167–174, and 177–180. None of these isotopes has any commercial use.

Extraction

Lutetium is the most difficult lanthanide to obtain in pure form. The usual method used begins with either lutetium fluo-

ride (LuF_3) or lutetium chloride ($LuCl_3$). An active metal, such as sodium (Na) or potassium (K) is then added to LuF_3 or $LuCl_3$ to obtain pure lutetium. For example:

$$LuF_3 + 3Na \rightarrow 3NaF + Lu$$

Uses

Lutetium is the most expensive lanthanide, selling for about $75 a gram. It is sometimes used as a catalyst in the petroleum industry. A catalyst is a substance used to speed up or slow down a chemical reaction. The catalyst does not undergo any change itself during the reaction. There are virtually no other uses for lutetium.

Compounds

There are no commercially important lutetium compounds.

Health effects

The health effects of lutetium are not known. In such cases, the best advice is to treat the element as if it were very toxic.

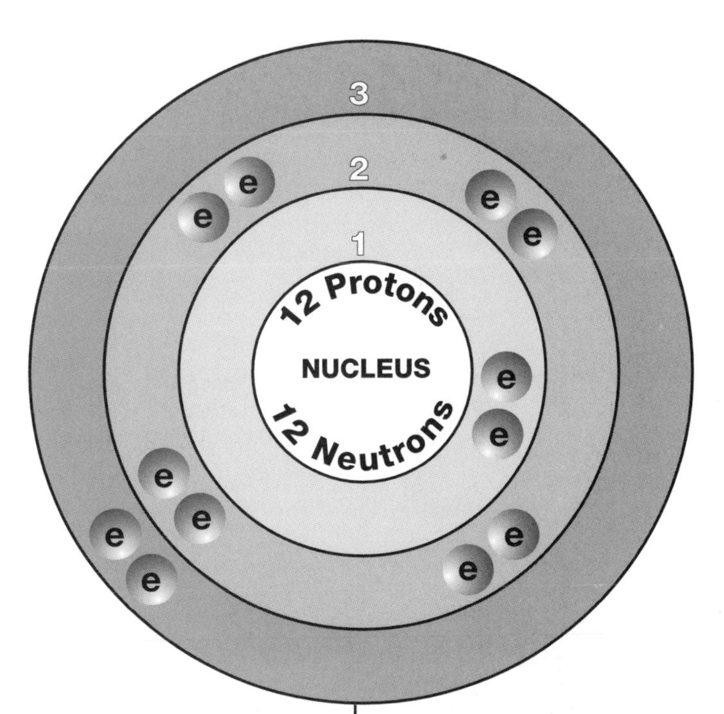

MAGNESIUM

Overview

Magnesium is the second element in Group 2 (IIA) of the periodic table a chart that shows how chemical elements are related to each other. The elements in Group 2 are known as the alkaline earth elements. Other elements in that group include **beryllium, calcium, strontium, barium,** and **radium.**

Compounds of magnesium have been used by humans for centuries. Yet, the element itself was not isolated until 1808. The long delay occurred because magnesium forms very stable compounds. That means that such compounds do not break down very easily.

Magnesium is the seventh most abundant element in the Earth's crust. It also occurs in large amounts dissolved in ocean waters.

Large amounts of magnesium are used to make alloys. An alloy is made by melting or mixing two or more metals. The mixture has properties different from those of the individual metals. Magnesium alloys are quite light, yet very strong. This property makes them useful in the construction of airplanes and spacecraft.

SYMBOL
Mg

ATOMIC NUMBER
12

ATOMIC MASS
24.305

FAMILY
Group 2 (IIA)
Alkaline earth metal

PRONUNCIATION
mag-NEE-zee-um

The allure of Epsom salts

Perhaps the best know magnesium compound is magnesium sulfate ($MgSO_4$). It is popularly known as Epson salts.

One of the earliest stories about Epsom salts dates back to 1618. The town of Epsom, in Surrey, England, was suffering from a severe drought. A farmer named Henry Wicker brought his cattle to drink from a water hole on the town commons (central park). But the cattle would not drink the water. Wicker was surprised because he knew they were very thirsty. He tasted the water himself and found that it was very bitter.

The bitterness was due to magnesium sulfate in the water. This compound became known as Epsom salts.

People soon learned that soaking in the natural waters that contained Epsom salts made them feel better. The salts seemed to have properties that soothed the body. Before long, soaking in these waters became very popular.

Today, Epsom salts are used in bath water. They relax sore muscles and remove rough skin. Many people believe the salts have the same relaxing effect as hot springs. Some gardeners even believe that sprinkling Epsom salts in the garden helps flowers and vegetables grow!

About 70 percent of the magnesium compounds produced in the United States are used in the manufacture of refractory materials. A refractory material is one that can withstand very high temperatures by reflecting heat. Refractory materials are used to line the ovens that maintain high temperatures. The remaining 30 percent of magnesium compounds are used in agriculture, construction, industrial, and chemical operations.

Discovery and naming

Compounds of magnesium are very abundant in the Earth. Dolomite, or calcium magnesium carbonate ($CaMg(CO_3)_2$), is an example. Dolomite has been used as a building material for centuries.

Another well-known magnesium compound is Epsom salts, or magnesium sulfate ($MgSO_4$). Epsom salts are known for their soothing qualities, most notably when added to a bath. (See accompanying sidebar.)

Careful studies of magnesium and its compounds began in the middle 1700s. Scottish physician and chemist Joseph Black (1728–99) carried out some of the earliest experiments on magnesium compounds. He reported on his research in an article that became famous. Black is sometimes given credit for "discovering" magnesium because of his work with the element.

By 1800, chemists knew that magnesium was an element. But no one had been able to prepare pure magnesium metal. Magnesium holds very tightly to other elements in its compounds. No one had found a way to break the bonds between magnesium and these other elements.

In 1808, English chemist Humphry Davy (1778–1829) solved the problem by passing an electric current through molten (melted) magnesium oxide (MgO). The current caused the compound to break apart, forming magnesium metal and **oxygen** gas:

$$2MgO \xrightarrow{\text{electric current}} 2Mg + O_2$$

Davy used this method to discover a number of other elements. (See sidebar on Davy in the **calcium** entry in Volume 1.) Like magnesium, these elements form compounds that are very difficult to break apart. An electric current provides the energy to break these compounds down into their elements.

The name magnesium goes back many centuries. It was selected in honor of a region in Greece known as Magnesia. The region contains large supplies of magnesium compounds.

Physical properties

Magnesium is a moderately hard, silvery-white metal. It is the lightest of all structural metals. These metals are strong enough to be used to build buildings, bridges, automobiles, and airplanes.

Magnesium is easily fabricated. Fabrication means shaping, molding, bending, cutting, and working with a metal. Metals must be fabricated before they can be turned into useful products. Metals that are strong, tough, or hard are not easily fabricated. They must be converted to an alloy. A metal that *is* more easily fabricated (such as magnesium) is combined with them.

The melting point of magnesium is 651°C (1,200°F) and its boiling point is 1,100°C (2,000°F). Its density is 1.738 grams per cubic centimeter.

WORDS TO KNOW

Alkaline earth metal an element found in Group 2 (IIA) of the periodic table

Alloy a mixture of two or more metals with properties different from those of the individual metals

Enzyme catalyst in a living organism that speeds up the rate at which certain changes take place in the body

Fabrication shaping, molding, bending, cutting, and working with a metal

Isotopes two or more forms of an element that differ from each other according to their mass number

Periodic table a chart that shows how the chemical elements are related to each other

Radioactive isotope an isotope that breaks apart and gives off some form of radiation

Refractory a material that can withstand very high temperatures and reflects heat from itself

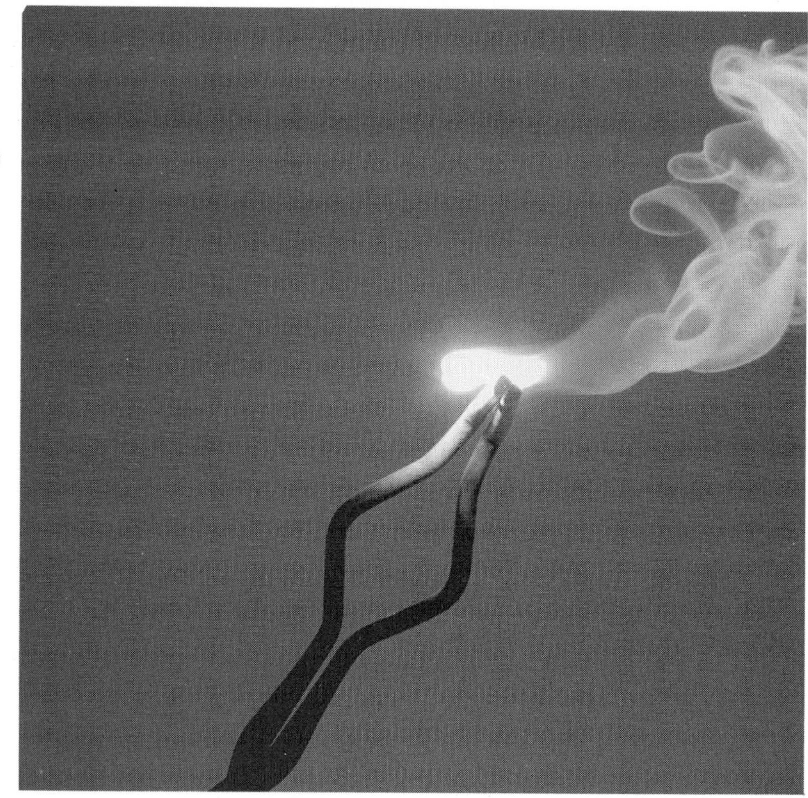

At high temperatures, magnesium burns with a blinding white light.

Chemical properties

Magnesium is a fairly active metal. It reacts slowly with cold water and more rapidly with hot water. It combines with oxygen at room temperature to form a thin skin of magnesium oxide. It burns with a blinding white light at higher temperatures. Magnesium reacts with most acids and with some alkalis. An alkali is a chemical with properties opposite those of an acid. Sodium hydroxide (common lye such as Drano) and limewater are examples of alkalis.

Magnesium also combines easily with many non-metals, including **nitrogen, sulfur, phosphorus, chlorine, fluorine, bromine,** and **iodine.** It also reacts readily with a number of compounds, such as carbon monoxide (CO), carbon dioxide (CO_2), sulfur dioxide (SO_2), and nitric oxide (NO).

Occurrence in nature

The abundance of magnesium in the Earth's crust is estimated to be about 2.1 percent. That makes it the sixth most common element in the earth. It also occurs in seawater. A cubic mile of seawater is estimated to contain up to six million tons of magnesium.

There are many naturally occurring minerals of magnesium. Some of the most important are dolomite; magnesite, or magnesium carbonate ($MgCO_3$); carnallite, or potassium magnesium chloride ($KMgCl_3$); and epsomite, or magnesium sulfate ($MgSO_4$).

The largest producer of magnesium ores is Turkey. Other large producers are North Korea, China, Slovakia, Austria, and Russia. The amount of magnesium produced in the United States is not reported in order to protect trade secrets.

Magnesium produced in the United States comes from three sources: seawater, brine, and mines. Seawater is processed to obtain magnesium by companies in California, Delaware, Florida, and Texas. Magnesium is obtained from brine in Michigan and Utah. Brine is water that is even saltier than seawater. Finally, some magnesium compounds are taken from mines in Nevada, North Carolina, and Washington.

Isotopes

There are three naturally occurring isotopes of magnesium: magnesium-24, magnesium-25, and magnesium-26. Isotopes are two or more forms of an element. Isotopes differ from each other according to their mass number. The number written to the right of the element's name is the mass number. The mass number represents the number of protons plus neutrons in the nucleus of an atom of the element. The number of protons determines the element, but the number of neutrons in the atom of any one element can vary. Each variation is an isotope.

One radioactive isotope of magnesium, magnesium-28, also exists. A radioactive isotope is one that breaks apart and gives off some form of radiation. Radioactive isotopes are produced when very small particles are fired at atoms. These particles stick in the atoms and make them radioactive. Magnesium-28 has no important commercial uses.

Extraction

Magnesium is prepared by one of two methods. The first method is similar to the method used by Davy in 1808. An electric current is passed through molten (melted) magnesium chloride:

Magnesium alloys are often used in the production of skis.

$$2MgCl \xrightarrow{\text{electric current}} 2Mg + Cl_2$$

The second method involves reacting magnesium oxide with ferrosilicon. Ferrosilicon is an alloy of **iron** and **silicon.** When magnesium oxide and ferrosilicon react, free magnesium metal is formed.

Uses

Although most cameras now use electronic flashes, magnesium metal is often contained in cameras that use flash bulbs. A thin strip of magnesium metal is inside the bulb. When the

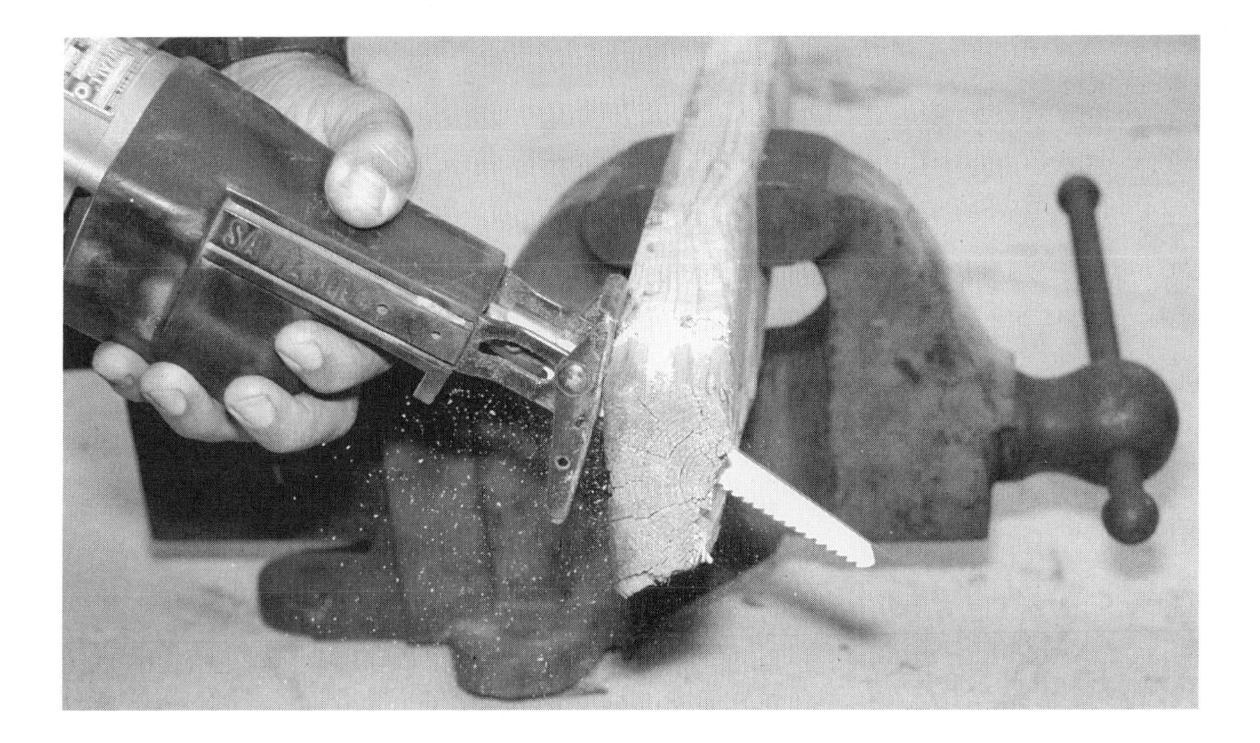

Magnesium alloys are used in power tools, such as the reciprocating saw shown here.

flash is ignited, the magnesium strip catches fire. It burns with a very bright white light. The light from the bulb illuminates a scene for the photograph.

A common use of magnesium metal is in fireworks. Most firework displays include some brilliant flashes of very white light. Those flashes are produced by the burning of magnesium metal.

Magnesium is commonly alloyed with other metals. Magnesium and aluminum, for instance, are two metals that combine to form alloys that are very strong and resistant to corrosion (rust). But they weigh much less than steel alloys with similar properties.

Strength and low density are important properties in the manufacture of airplanes, automobiles, metal luggage, ladders, shovels and other gardening equipment, racing bikes, skis, race cars, cameras, and power tools. A typical magnesium alloy contains about 90 percent magnesium, 2 to 9 percent aluminum, and small amounts of **zinc** and manganese.

Compounds

The largest single use of magnesium compounds is in refractories. Other magnesium compounds are used in the following categories:

medicine: pain killer and fever reducer (magnesium acetylsalicylate); antacid to neutralize stomach acid (magnesium hydroxide; magnesium phosphate; magnesium silicate); laxative to loosen the bowels (magnesium carbonate; magnesium chloride; magnesium citrate; magnesium hydroxide; magnesium lactate; magnesium phosphate); antiseptic to kill germs (magnesium borate; magnesium salicylate; magnesium sulfate); sedative to help one get sleep (magnesium bromide)

production of glass and ceramics: magnesium fluoride; magnesium oxide

mothproofing of textiles: magnesium hexafluorosilicate

fireproofing wood for construction: magnesium phosphate

manufacture of paper: magnesium sulfite

Magnesium forms part of the chlorophyll molecule found in all green plants.

Health effects

Magnesium is essential for good health in both plants and animals. It forms part of the chlorophyll molecule found in all green plants. Chlorophyll is the molecule in green plants that controls the conversion of carbon dioxide and water to carbohydrates, such as starch and sugar. Plants that do not get enough magnesium cannot make enough chlorophyll. Their leaves develop yellowish blotches as a result.

Magnesium is found in many enzymes in both plants and animals. An enzyme is a catalyst in a living organism. It speeds up the rate at which certain changes take place in the body. Enzymes are essential in order for living cells to function properly. It is difficult *not* to get enough magnesium in one's daily diet. It is found in nuts, cereals, seafoods, and green vegetables. Most people have no problem getting the 300 to 400 milligrams of magnesium recommended in the daily diet.

A lack of magnesium can occur, however. For example, alcoholics and children in poor countries sometimes develop a magnesium deficiency. In such cases, magnesium deficiency may cause a person to become easily upset or overly aggressive.

On the other hand, it is also possible to be exposed to too much magnesium. For example, inhaling magnesium powder can produce irritation of the throat and eyes, resulting in a fever. In large doses, magnesium can cause damage to muscles and nerves. It can eventually result in loss of feeling and paralysis (inability to move parts of the body).

Such conditions are rare. They are likely to occur only among people who have to work with magnesium metal on a regular basis.

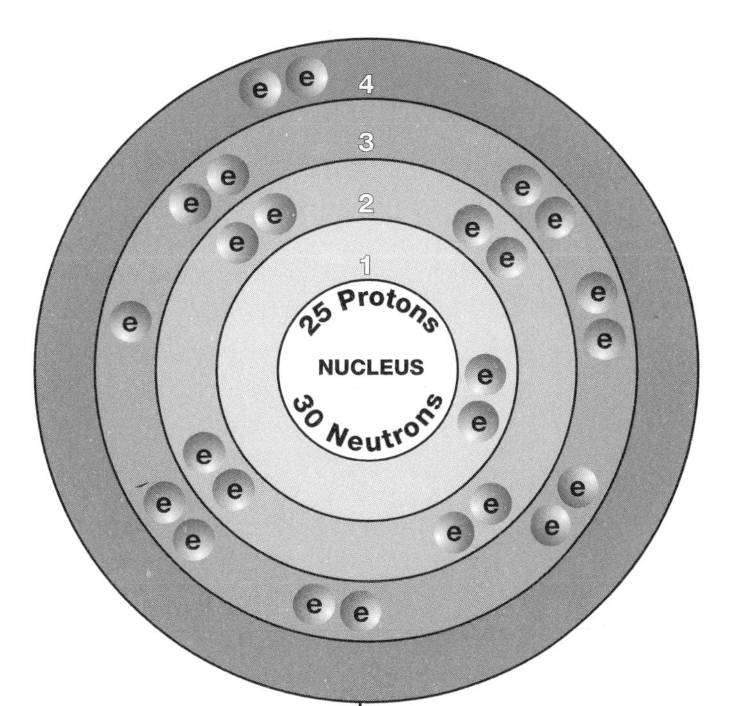

MANGANESE

Overview

Manganese is a transition metal. The transition metals are the large block of elements in the middle of the periodic table. The periodic table is a chart that shows how chemical elements are related to each other. The transition metals make up Rows 4 through 7 in Groups 3 through 12 of the periodic table. Many of the best known and most widely used metals are in this group of elements.

It took chemists some time to discover the difference between manganese and **iron.** The two metals have very similar properties and often occur together in the Earth's crust. The first person to clearly identify the differences between the two elements was Swedish mineralogist Johann Gottlieb Gahn (1745–1818) in 1774.

Manganese plays an interesting role in the U.S. economy. It is absolutely essential in the production of iron and steel. No element has been found that can replace manganese is such applications. The United States has essentially no manganese supplies of its own, so it depends on imports from other nations.

SYMBOL
Mn

ATOMIC NUMBER
25

ATOMIC MASS
54.9380

FAMILY
Group 7 (VIIB)
Transition metal

PRONUNCIATION
MANG-guh-neez

Discovery and naming

One of the main ores of manganese is pyrolusite. Pyrolusite is made up primarily of the compound manganese dioxide (MnO_2). Early artists were familiar with pyrolusite. They used the mineral to give glass a beautiful purple color. They also used the mineral to remove color from a glass. When glass is made, it often contains impurities that give the glass an unwanted color. The presence of iron, for example, can give glass a yellowish tint. Adding pyrolusite to yellowish glass removes the color. The purple tint of pyrolusite balances out the yellow color of the glass. The glass ends up being clear and colorless.

By the mid-1700s, chemists began to suspect that pyrolusite might contain a new element. Some authorities credit German chemist Ignatius Gottfried Kaim with isolating the element in 1770. However, Kaim's report was not read by many chemists and was quickly lost.

During this period, some of the most famous chemists in Europe were trying to analyze pyrolusite, but none of them was successful. The problem was solved in 1774 when Gahn developed a method for removing the new element from pyrolusite. He heated pyrolusite with charcoal (pure **carbon**). The carbon took **oxygen** away from manganese dioxide, leaving behind pure manganese:

$$MnO_2 + C \rightarrow CO_2 + Mn$$

The origin of manganese's name is a bit confusing. Early chemists associated the new element with a mineral called magnesia. That mineral got its name because it is magnetic. Magnesia does not contain manganese, but the name stuck.

Physical properties

Manganese is a steel-gray, hard, shiny, brittle metal. It is so brittle, in fact, that it cannot be machined in its pure form. Machining refers to the bending, cutting, and shaping of a metal by mechanical means. The melting point of manganese is 1,245°C (2,273°F) and its boiling point is about 2,100°C (3,800°F). Its density is 7.47 grams per cubic centimeter.

Manganese exists in four allotropic forms. Allotropes are forms of an element with different physical and chemical properties.

WORDS TO KNOW

Allotropes forms of an element with different physical and chemical properties

Alloy a mixture of two or more metals with properties different from those of the individual metals

Isotopes two or more forms of an element that differ from each other according to their mass number

Machining the bending, cutting, and shaping of a metal by mechanical means

Periodic table a chart that shows how the chemical elements are related to each other

Radioactive isotope an isotope that breaks apart and gives off some form of radiation

Transition metal an element in Groups 3 through 12 of the periodic table

The element changes from one form to another as the temperature rises. The form that exists from room temperature up to about 700°C (1,300°F) is the most common form.

Chemical properties

Manganese is a moderately active metal. It combines slowly with oxygen in the air to form manganese dioxide (MnO_2). At higher temperatures, it reacts more rapidly. It may even burn, giving off a bright white light. Manganese reacts slowly with cold water, but more rapidly with hot water or steam. It dissolves in most acids with the release of hydrogen gas. It also combines with **fluorine** and **chloride** to make manganese difluoride (MnF_2) and manganese dichloride ($MnCl_2$).

Occurrence in nature

Manganese never occurs as a pure element in nature. It always combines with oxygen or other elements. The most common ores of manganese are pyrolusite, manganite, psilomelane, and rhodochrosite. Manganese is also found mixed with iron ores. The largest producers of manganese ore in the world are China, South Africa, the Ukraine, Brazil, Australia, Gabon, and Kazakstan.

Manganese also occurs abundantly on the ocean floor in the form of nodules. These nodules are fairly large lumps of metallic ores. They usually contain **cobalt, nickel, copper,** and iron, as well as manganese. Scientists estimate that up to 1.5 trillion metric tons of manganese nodules may lie on the floors of the world's oceans and large lakes. Currently, there is no profitable method for removing these ores.

Manganese is the 12th most abundant element in the Earth's crust. Its abundance is estimated to be 0.085 to 0.10 percent. That makes it about as abundant as fluorine or **phosphorus.**

Isotopes

Only one naturally occurring isotope of manganese exists, manganese-22. Isotopes are two or more forms of an element. Isotopes differ from each other according to their mass number. The number written to the right of the element's name is the mass number. The mass number represents the number of protons plus neutrons in the nucleus of an atom of the element. The number of protons determines the element, but the number of neutrons in the atom of any one element can vary. Each variation is an isotope.

Up to 1.5 trillion metric tons of manganese nodules (large lumps of metallic ores) may lie on ocean floors.

Men and machine lay railroad tracks. A common alloy of manganese, ferromanganese, is contained in the steel used to produce railroad tracks.

Nine radioactive isotopes of manganese are known also. A radioactive isotope is one that breaks apart and gives off some form of radiation. Radioactive isotopes are produced when very small particles are fired at atoms. These particles stick in the atoms and make them radioactive.

None of the radioactive isotopes of manganese has any important commercial uses.

Extraction

The usual method for producing pure manganese is to heat manganese dioxide (MnO_2) with carbon or aluminum. These elements remove the oxygen and leave pure metal:

$$MnO_2 + C \rightarrow Mn + CO_2$$

Uses

Up to 90 percent of all manganese produced is made into steel alloys. An alloy is made by melting and mixing two or more metals. The mixture has properties different from those of the individual metals. The addition of manganese to steel makes the final product hard, as well as resistant to corrosion (rusting) and mechanical shock.

The heavy steel found in bank vaults contains ferromanganese, a manganese alloy.

The most common alloy of manganese is ferromanganese. This alloy contains about 48 percent manganese combined with iron and carbon. Ferromanganese is the starting material for making a very large variety of steel products, including tools, heavy-duty machinery, railroad tracks, bank vaults, construction components, and automotive parts. About 60 percent of the manganese used in the United States in 1996 went to the manufacture of ferromanganese.

Another common alloy of manganese is silicomanganese. It contains manganese, **silicon,** and carbon in addition to iron. It is used for structural components and in springs. The production of silicomanganese accounted for about 33 percent of the manganese used in the United States in 1996.

Manganese is also used to make alloys with metals other than iron or steel. For example, the alloy known as manganin is 84 percent copper, 12 percent manganese, and 4 percent nickel. Manganin is used in electrical instruments.

Compounds

Less than 10 percent of all the manganese used in the United States goes to the production of manganese compounds. Perhaps the most important commercial use of these compounds is manganese dioxide (MnO_2). Manganese dioxide is used to make dry-cell batteries. These batteries are used in electronic equipment, flashlights, and pagers. Dry cell batteries hold a black pasty substance containing manganese dioxide. The use of manganese dioxide in a dry cell prevents hydrogen gas from collecting in the battery as electricity is produced.

Another manganese compound, manganous chloride ($MnCl_2$), is an additive in animal food for cows, horses, goats, and other domestic animals. Fertilizers also contain manganous chloride so that plants get all the manganese they need.

Manganous chloride ($MnCl_2$) is an additive in animal food.

Finally, small amounts of manganese compounds are used as coloring agents in bricks, textiles, paints, inks, glass, and ceramics. Manganese compounds can be found in shades of pink, rose, red, yellow, green, purple, and brown.

Health effects

Manganese is one of the chemical elements that has both positive and negative effects on living organisms. A very small amount of the element is needed to maintain good health in plants and animals. The manganese is used by enzymes in an organism. An enzyme is a molecule that makes chemical reactions occur more quickly in cells. Enzymes are necessary to keep any cell operating properly. If manganese is missing from the diet, enzymes do not operate efficiently. Cells begin to die, and the organism becomes ill.

Fortunately, the amount of manganese needed by organisms is very small. It is not necessary to take extra manganese to meet the needs of cells.

In fact, an excess of manganese can create health problems. These problems include weakness, sleepiness, tiredness, emotional disturbances, and even paralysis. The only way to receive such a large dose is in a factory or mine. Workers may inhale manganese dust in the air.

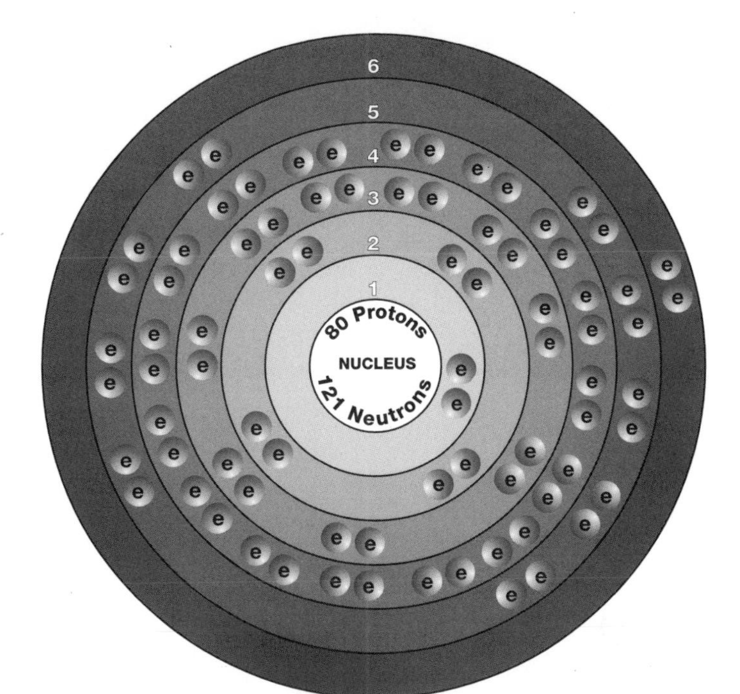

MERCURY

Overview

Mercury is a transition metal. A transition metal is one of the elements found between Groups 2 (IIA) and 13 (IIIA) on the periodic table. The periodic table is a chart that shows how chemical elements are related to one another. Mercury has long been known as quicksilver, because it is a silver liquid. The chemical symbol also reflects this property. The symbol, Hg, comes from the Latin term *hydrargyrum,* meaning "watery silver."

Mercury has been known for thousands of years. In many cultures, people learned to make mercury metal from its most important ore, cinnabar. When heated cinnabar releases mercury as a vapor (gas). The vapor is cooled and captured as liquid mercury.

Some mercury compounds are known to be poisonous. For example, mercuric chloride (corrosive sublimate) was often used to kill pests and, sometimes, people. On the other hand, some mercury compounds have been used as medicines. For instance, mercurous chloride (calomel) was long used as a cure for skin rashes. In the last forty years, the dangers of mercury have become better known. As a result, mercury use is now being phased out.

SYMBOL
Hg

ATOMIC NUMBER
80

ATOMIC MASS
200.59

FAMILY
Group 12 (IIB)
Transition metal

PRONUNCIATION
MER-kyuh-ree

WORDS TO KNOW

Amalgam a combination of mercury with at least one other metal

Distillation a process by which two or more liquids can be separated from each other by heating them to their boiling points

Isotopes two or more forms of an element that differ from each other according to their mass number

Periodic table a chart that shows how the chemical elements are related to each other

Phosphor a material that gives off light when struck by electrons

Radioactive isotope an isotope that breaks apart and gives off some form of radiation

Surface tension a property of liquids that makes them act like they are covered with a skin

Toxic poisonous

Transition metal an element in Groups 3 through 12 of the periodic table

Discovery and naming

The oldest sample of mercury dates to about the fifteenth or sixteen century B.C. It was found in an Egyptian tomb at Kurna, stored in a small glass container.

Mercury and cinnabar are both mentioned in ancient manuscripts. The Chinese, Hindus, Egyptians, Greeks, and Romans all recorded information about the element and its ore. Greek philosopher Theophrastus (372–287 B.C.), for example, described a method for preparing mercury. Cinnabar was rubbed together with vinegar in a clay dish. Theophrastus wrote that the cinnabar had been found in **silver** mines. When the metal was first made, he said, people thought it might contain **gold.** They were misled by the metal's shiny appearance. They soon realized, however, that it was quite different from gold.

Many reports on mercury told of its poisonous effects. Slaves who worked in Roman mercury mines, for example, often died of exposure to mercury. Strangely enough, trees and plants around these mines were not affected. Mercury was sometimes very dangerous and sometimes quite safe. People even drank from streams that ran through mercury mines. Scientists now know that mercury's effects depend on the form in which it occurs.

Mercury amalgams have also been around for a long time. An amalgam is a combination of mercury with at least one other metal. Amalgams are formed when a metal, such as silver, dissolves in mercury. The process is similar to dissolving salt in water. Amalgamation is used in mining to remove silver from ore. The silver dissolves in the mercury and a silver amalgam is formed. Heating the amalgam releases the silver. This method was used by miners as early as the sixteenth century.

Physical properties

Mercury is the only liquid metal. In fact, there is only one other liquid element, **bromine.** Bromine is a non-metal. Mercury can be frozen (changed into a solid) at a temperature of −38.85°C (−37.93°F). It can be changed into a gas ("boiled") at 365.6°C (690.1°F). Its density is 13.59 grams per cubic centimeter.

Mercury has two physical properties of special interest. First, it has very high surface tension. Surface tension is a property of liquids that make them act like they are covered with a skin.

Droplets of mercury, the only liquid metal.

For example, some water bugs are able to walk on the surface of water. With care, one can float a needle on the surface of water. These incidents are possible because of water's surface tension.

Mercury is also a very good conductor of electricity. This property is used in a number of practical devices. One such device is a mercury switch, such as the kind that turns lights on and off. A small amount of mercury can be placed into a tiny glass capsule. The capsule can be made to tip back and forth. As it tips, the mercury flows from one end to the other. At one end

of the capsule, the mercury may allow an electric current to flow through a circuit. At the other end, no mercury is present, so no current can flow. Mercury switches are easy to make and very efficient.

Chemical properties

Mercury is moderately active. It does not react with **oxygen** in the air very readily. It reacts with some acids when they are hot, but not with most cold acids.

Occurrence in nature

Mercury is the only liquid metal.

The abundance of mercury in the Earth's crust is estimated to be about 0.5 parts per million. That makes it one of the 20 least common elements. It very rarely occurs as an element. Instead, it is usually found as a compound. Its most common ore is cinnabar, or mercuric sulfide (HgS). Cinnabar usually occurs as a dark red powder. It is often called by the common name of vermillion or Chinese vermillion.

The largest producer of mercury outside the United States is Spain. U.S. production numbers are not announced in order to protect U.S. industries from revealing important company secrets. Other producers after Spain are Kyrgyzstan, Algeria, China, and Finland.

In the United States, mercury is produced as a by-product of gold mining. It comes from eight gold mines in California, Nevada, and Utah.

Isotopes

Seven naturally occurring isotopes of mercury are known. They are mercury-196, mercury-198, mercury-199, mercury-200, mercury-201, mercury-202, and mercury-204. Isotopes are two or more forms of an element. Isotopes differ from each other according to their mass number. The number written to the right of the element's name is the mass number. The mass number represents the number of protons plus neutrons in the nucleus of an atom of the element. The number of protons determines the element, but the number of neutrons in the atom of any one element can vary. Each variation is an isotope.

About a dozen radioactive isotopes of mercury are known also. A radioactive isotope is one that breaks apart and gives off

some form of radiation. Radioactive isotopes are produced when very small particles are fired at atoms. These particles stick in the atoms and make them radioactive.

Two radioactive isotopes of mercury are used in medicine, mercury-197 and mercury-203. Both isotopes are used to study the brain and the kidneys. The isotopes are injected into the body where they travel to the brain and the kidneys. Inside these two organs, the isotopes give off radiation that is detected by instruments held above the body. The pattern of radiation provides information about how well the brain and kidneys are functioning.

Extraction

Mercury is still prepared as it was hundreds of years ago. Cinnabar is heated in air. The compound breaks down to give mercury metal:

$$HgS \xrightarrow{\text{heated}} Hg + S$$

The mercury metal is then purified by distillation. Distillation is the process of heating two or more liquids to their boiling points. Different liquids boil at different temperatures. The liquid that is wanted (such as mercury) can be collected at *its* boiling point. Mercury that is more than 99 percent pure can be collected by distillation.

Uses

The most important use of mercury is in the preparation of chlorine. Chlorine is produced by passing an electric current through sodium chloride:

$$2NaCl \xrightarrow{\text{electric current}} 2Na + Cl_2$$

There is a problem with using this method, however. Sodium (Na) is a very reactive metal. If any water is present, the sodium will react violently with the water. This reaction makes the production of chlorine much more difficult.

In 1892, two English chemists developed a method for solving this problem. They made a container with a layer of mercury on the bottom. As sodium is produced by the electric current, it dissolves in the mercury, forming an amalgam. The sodium is unable to react with water. For many years, the "mercury cell"

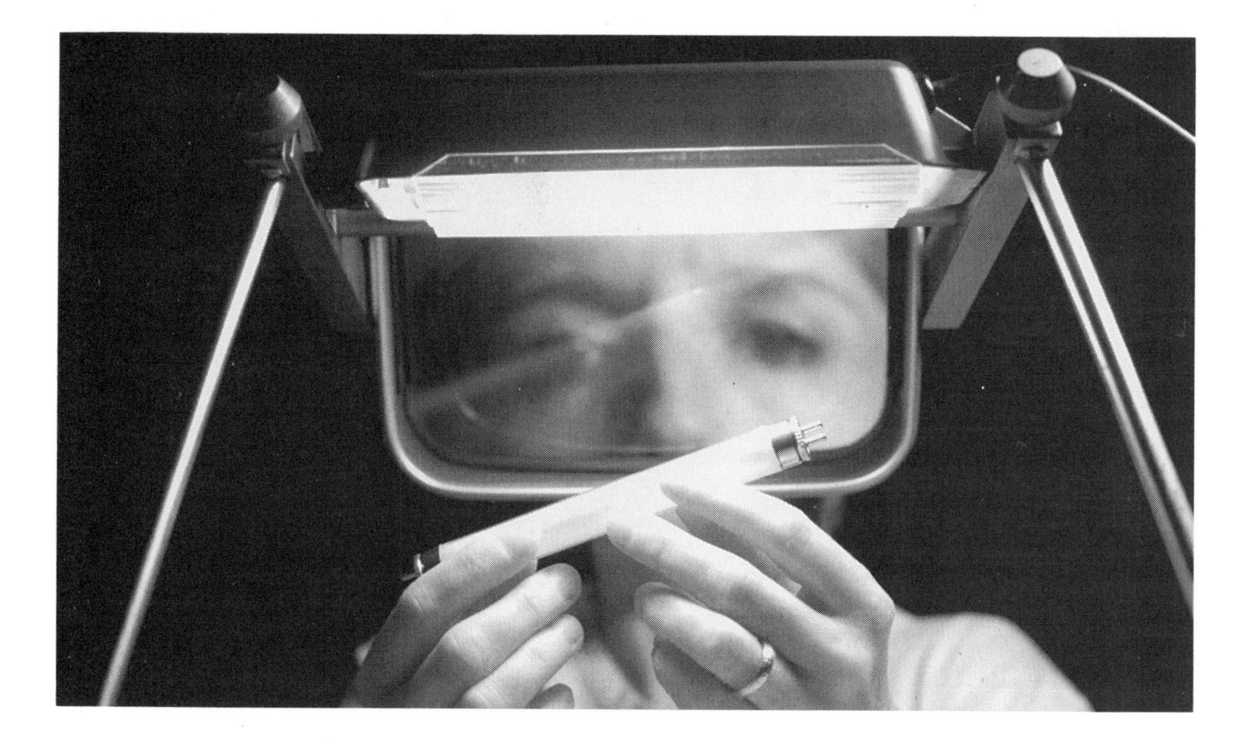

With fluorescent lights, when an electric current passes through mercury vapors, the resulting invisible radiation strikes phosphors. This creates visible light.

invented in 1892 was a very popular method for producing chlorine.

But today, companies are looking for other ways to make chlorine. They are worried about the harmful effects of mercury. They are also concerned that mercury can get into the environment and harm humans, animals, and plants.

The second most important use of mercury in the United States is in switches and other electrical applications. Again, there are increasing concerns about the health effects of mercury. Many companies are switching to electronic switches.

One application in which concerns about mercury have had little effect is fluorescent lamps. A fluorescent lamp contains mercury vapor (gas). When the lamp is turned on, an electric current passes through the mercury vapor, causing it to give off invisible radiation. The radiation strikes the inside of the glass tube, whose walls are coated with a phosphor. A phosphor is a material that gives off visible light when struck by electrons. The tube glows as the radiation strikes the phosphor.

Lamp manufacturers have reduced the amount of mercury in fluorescent lamps by about 60 percent. They developed ways to make the lamps work just as well with less mercury. However, mercury lamps are much more popular. Each lamp now contains much less mercury. But there are many more lamps than ever before.

For a time, mercury batteries were quite popular. In the early 1980s, more than 1,000 tons of mercury a year were used to make mercury batteries. These batteries are a special environmental problem, however. People tend to just throw them away when they no longer work. The cases split open easily, releasing mercury into the environment. As a result, much less mercury is now being used to make such batteries. In 1996, less than one ton of mercury was used in these batteries. They are now restricted almost entirely to military and medical uses.

Mercury is also used in dental applications, measuring instruments (such as mercury thermometers and barometers), and coatings for mirrors.

Compounds

Mercury compound use is also decreasing because of health concerns. A few of the compounds still in use follow. Notice that two different endings are used for mercury compounds. Those that end in *-ous* have less mercury than those that end in *-ic.*

mercuric arsenate ($HgHAsO_4$): waterproofing paints

mercuric benzoate ($Hg(C_7G_5O_2)_2$): medicine; used to treat syphilis

mercuric chloride, or mercury bichloride, or corrosive sublimate ($HgCl_2$): disinfectant, tanning of leather, spray for potato seedlings (to protect from disease), insecticide, preservation of wood, embalming fluid, textile printing, and engraving

mercuric cyanide ($Hg(CN)_2$): germicidal soaps (soaps that kill germs), photography

mercuric oxide (HgO): red or yellow pigment in paints, disinfectant, fungicide (to kill fungi), perfumes and cosmetics

The tragic effects of mercury poisoning

In a tragic irony, a scientist who was helping to improve the environment died as a result of her efforts. On June 8, 1997, Dartmouth College chemistry professor Karen Wetterhahn died of mercury poisoning. Less than a year earlier, she had been experimenting with dimethyl mercury when she spilled a tiny amount on her hands. Dimethyl mercury is one of the most toxic of mercury compounds.

Wetterhahn was studying the effects that heavy metals (mercury, chromium, lead, and arsenic) have on living things. She was concerned about how these elements pollute the environment and cause disease in people.

In August 1996, as Wetterhahn was transferring some dimethyl mercury to a tube, the accident occurred. She was wearing latex gloves, but they were not adequate protection against the dangerous chemical. The mercury seeped into her skin. Wetterhahn did not begin to feel the effects of the exposure until six months later. She then started losing her balance, slurring her speech, and suffering vision and hearing loss. Tests showed her system had eighty times the lethal dose of mercury. Wetterhahn died of mercury poisoning on June 8, 1997.

Wetterhahn's death prompted some safety changes. Bright stickers on latex glove boxes should warn against using the gloves with hazardous chemicals. Workshops were held to teach proper glove selection. The dangers of dimethyl mercury were stressed. And scientists were urged to use a less dangerous chemical than dimethyl mercury. Overall, her death heightened awareness in the scientific community of potential laboratory dangers.

mercuric sulfide (HgS): red or black pigment in paints

mercurous chloride, or calomel (Hg_2Cl_2): fungicide, maggot control in agriculture, fireworks

mercurous chromate (Hg_2CrO_4): green pigment in paints

mercurous iodide (Hg_2I_2): kills bacteria on the skin

Health effects

Mercury metal and most compounds of mercury are highly toxic. Interestingly enough, scientists have become aware of this fact only quite recently. The toxicity of *some* mercury compounds has been known for many centuries. One form of mercury chloride known as calomel, for example, was sometimes used as a poison to kill people. It was also once used extensively to kill fungi and control maggots in agricultural crops.

"Mad as a hatter!"

Back in the 1800s, most of the negative effects of mercury and its compounds were not yet known. Hatmakers of that time commonly used a mercury compound in their craft. It was used to treat the felt and beaver fur that lined the hats. Eventually, exposure to the mercury began to cause changes in the hatmakers' bodies. Their personalities and behavior became erratic. Recognizing the bizarre personalities of many hatmakers, people often used the expression "mad as a hatter." In fact, author Lewis Carroll (1832–98) created a character for *Alice's Adventures in Wonderland* that owes its origins to the symptoms of mercury poisoning: The Mad Hatter.

But even as recently as fifty years ago, there was relatively little concern about mercury metal and many mercury compounds. High school chemistry students often played with tiny droplets of mercury in the laboratory. They used mercury to coat pennies and other pieces of metal.

Mercury was also widely used in dentistry. It was used to make amalgams, alloys of mercury with other metals, used to fill teeth. Most people even today are likely to have dental fillings that contain a small amount of mercury metal.

In the last fifty years, chemists have learned a great deal more about the toxic effects of both mercury metal and most of its compounds. They now know that mercury itself enters the body very easily. Its vapors pass through the skin into the blood stream. Its vapors can also be inhaled. And, of course, it can also be swallowed. In any of these cases, mercury gets into blood and then into cells. There it interferes with essential chemical reactions and can cause illness and death.

Sometimes, these effects occur over very long periods of time. People who work with mercury, for example, may take in small amounts of mercury over months or years. Health problems develop very slowly. These problems can include inflammation of the mouth and gums; loosening of the teeth; damage to the kidneys and muscles; shaking of the arms and legs; and depression, nervousness, and personality changes.

People can also be exposed to large doses of mercury over short periods of time. In such cases, even more serious health problems can arise. These include nausea, vomiting diarrhea, stomach pain, damage to the kidneys, and death in only a week or so.

So is mercury still safe to use in dental fillings? That question is the source of considerable controversy. Some people say that so little mercury is lost from fillings that the metal presents no danger to people. Other people think that dentists should take no chances with this dangerous metal. They should stop using mercury fillings entirely.

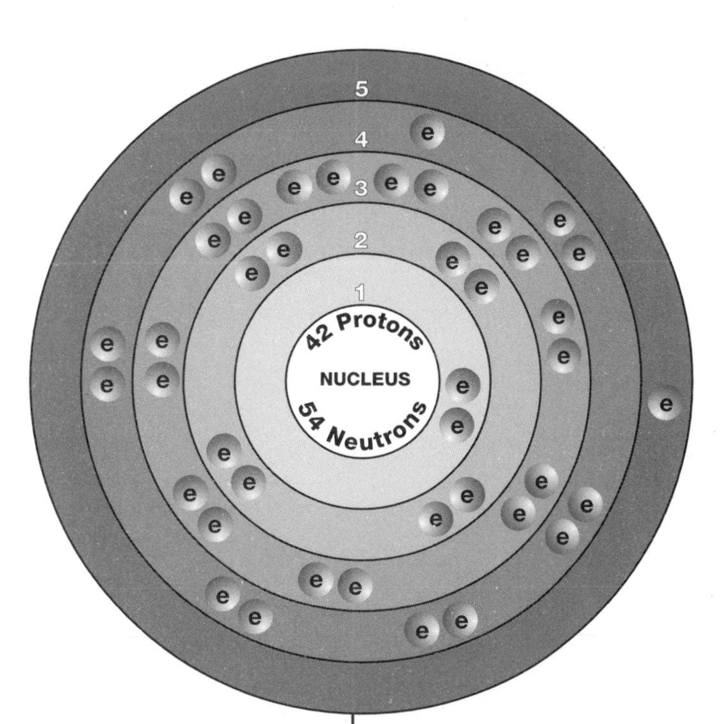

MOLYBDENUM

Overview

Molybdenum was one of the first metals to be discovered by a modern chemist. It was found in 1781 by Swedish chemist Peter Jacob Hjelm (1746–1813). Hjelm's work on the element was not published, however, until more than a century later.

Molybdenum is a transition metal, placing it in the center of the periodic table. The periodic table is a chart that shows how chemical elements are related to one another.

Molybdenum is a hard, silvery metal with a very high melting point. It is used primarily to make alloys with other metals. An alloy is a mixture of two or more metals. The mixture has properties different from those of the individual metals. The most common alloys of molybdenum are those with steel. Molybdenum improves the strength, toughness, resistance to wear and corrosion, and ability to harden steel.

SYMBOL
Mo

ATOMIC NUMBER
42

ATOMIC MASS
95.94

FAMILY
Group 6 (VIB)
Transition metal

PRONUNCIATION
muh-LIB-duh-num

Discovery and naming

The most common ore of molybdenum is called molybdenite. Molybdenite contains a compound of molybdenum and **sulfur,** molybdenum disulfide (MoS_2). Molybdenum disulfide is a soft black powder that looks like graphite. Graphite is pure **carbon;**

it makes up the "lead" in ordinary pencils. In fact, earlier chemists thought that graphite and molybdenum disulfide were the same material.

The soft "squishy" character of molybdenum disulfide frustrated early researchers of the compound. Chemists often grind up a material before trying to dissolve it in acids or other liquids. But molybdenum disulfide cannot be ground up. The material just slides out of the way.

It was not until 1781 that Hjelm found a way to work with the compound. He discovered that it was very different from graphite. In fact, he found that it contained an entirely new element. The name chosen for the new element illustrates a further confusion. In Greek, the word for **lead** is *molybdos*. The name chosen for the new element, molybdenum, is actually the Greek word for lead!

Hjelm's work was known to his fellow chemists because of letters they had written each other. But the report of his discovery was not actually printed for all chemists to read until 1890. Between 1791 and 1891, Hjelm's research was repeated by other chemists. They confirmed what he discovered, and he is recognized today as the discoverer of molybdenum.

Physical properties

As a solid, molybdenum has a silvery-white metallic appearance. It more commonly occurs as a dark gray or black powder with a metallic luster. Its melting point is about 2,610°C (about 4,700°F) and the boiling point is 4,800 to 5,560°C (8,600 to 10,000°F). Its density is 10.28 grams per cubic centimeter.

Chemical properties

Molybdenum does not dissolve in most common chemical reagents. A chemical reagent is a substance used to study other substances, such as an acid or an alkali. For example, molybdenum does not dissolve in hydrochloric acid, hydrofluoric acid, ammonia, sodium hydroxide, or dilute sulfuric acid. These chemicals are reagents often used to test how reactive a substance is. Molybdenum does dissolve in hot strong sulfuric or nitric acids, however. The metal does not react with **oxygen** at room temperatures, but does react with oxygen at high temperatures.

WORDS TO KNOW

Alloy a mixture of two or more metals that has properties different from those of the individual metals

Catalyst a substance used to speed up or slow down a chemical reaction without undergoing any change itself

Chemical reagent a substance, such as an acid or an alkali, used to study other substances

Isotopes two or more forms of an element that differ from each other according to their mass number

Radioactive isotope an isotope that breaks apart and gives off some form of radiation

Radioactive tracer an isotope whose movement in the body can be followed because of the radiation it gives off

Trace element an element that is needed in very small amounts for the proper growth of a plant or animal

Occurrence in nature

Molybdenum never occurs free in nature. Instead, it is always part of a compound. In addition to molybdenite, it occurs commonly as the mineral wulfenite ($PbMoO_4$). Its abundance in the Earth's crust is estimated to be about 1 to 1.5 parts per million. That makes it about as common as **tungsten** and many of the rare earth (lanthanide) elements. About two-thirds of all the molybdenum in the world comes from Canada, Chile, China, and the United States. In the United States, molybdenum ores are found primarily in Alaska, Colorado, Idaho, Nevada, New Mexico, and Utah.

Isotopes

Seven naturally occurring isotopes of molybdenum exist: molybdenum-92, molybdenum-94, molybdenum-95, molybdenum-96, molybdenum-97, molybdenum-98, and molybdenum-100. Isotopes are two or more forms of an element. Isotopes differ from each other according to their mass number. The number written to the right of the element's name is the mass number. The mass number represents the number of protons plus neutrons in the nucleus of an atom of the element. The number of protons determines the element, but the number of neutrons in the atom of any one element can vary. Each variation is an isotope.

Molybdenum disulfide is soft and squishy.

None of the seven naturally occurring molybdenum isotopes is radioactive. However, about a dozen artificial radioactive isotopes have been produced. A radioactive isotope is one that breaks apart and gives off some form of radiation. Radioactive isotopes are produced when very small particles are fired at atoms. These particles stick in the atoms and make them radioactive.

One radioactive isotope of molybdenum is commonly used in medicine, molybdenum-99m. (The "m" in this instance stands for "metastable," which means the isotope does not last very long.) This isotope is not used directly, however. Instead, it is used in hospitals to make another radioactive isotope technetium-99m. This isotope of **technetium** (atomic number 43) is widely used as a tracer for diagnostic studies of the brain, liver, spleen, heart, and other organs and body systems.

A radioactive tracer is an isotope whose movement in the body can be followed because of the radiation it gives off. The radi-

ation can be "traced" with special equipment held above the body. The pattern produced by the radiation allows a doctor to diagnose any unusual functioning (behavior) of the organ or body part.

Technetium-99m cannot be used for this purpose all by itself. It changes very quickly into a new isotope. Hospitals prepare molybdenum-99m first. This isotope can be stored for short periods of time. It slowly gives off radiation and changes into technetium-99m. The technetium-99m is captured as it is formed from molybdenum-99m and injected into the body for tracer studies. Because it is used to produced technetium-99m, the isotope molybdenum-99m is sometimes referred to as a "molybdenum cow."

Extraction

Pure molybdenum metal can be obtained from molybdenum trioxide (MoO_3) in a variety of ways. For example, hot hydrogen can be passed over the oxide to obtain the metal:

$$MoO_3 + 3H_2 \rightarrow 3H_2O + Mo$$

Uses

About 75 percent of the molybdenum used in the United States in 1996 was made into alloys of steel and iron. Nearly half of these alloys, in turn, were used to make stainless and heat-resistant steel. A typical use is in airplane, spacecraft, and missile parts. Another important use of molybdenum alloys is in the production of specialized tools. Spark plugs, propeller shafts, rifle barrels, electrical equipment used at high temperatures, and boiler plates are all made of molybdenum steel.

Another important use of molybdenum is in catalysts. A catalyst is a substance used to speed up or slow down a chemical reaction. The catalyst does not undergo any change itself during the reaction. Molybdenum catalysts are used in a wide range of chemical operations, in the petroleum industry, and in the production of polymers and plastics.

Compounds

A number of molybdenum compounds are used in industry and research. Interestingly, molybdenum disulfide is still used as a lubricant, as it was over two hundred years ago. The slippery black powder looks and behaves much like graphite. Molybde-

num is used in industrial operations to reduce the friction between sliding or rolling parts. It does not break down when heated or used for very long periods of time.

Other compounds of molybdenum are used as protective coatings in materials used at high temperatures; as solders; as catalysts; as additives to animal feeds; and as pigments and dyes in glasses, ceramics, and enamels.

Health effects

Molybdenum is relatively safe for humans and animals. No studies have shown it to be toxic. In fact, it is regarded as a necessary trace element for the growth of plants. A trace element is one that is needed in very small amounts for the proper growth of a plant or animal.

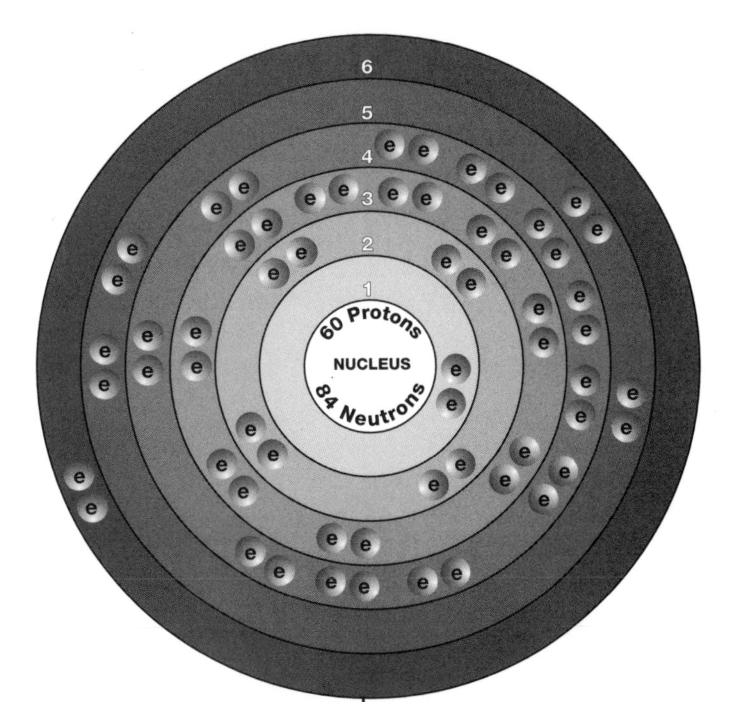

NEODYMIUM

SYMBOL
Nd

ATOMIC NUMBER
60

ATOMIC MASS
144.24

FAMILY
Lanthanide
(rare earth metal)

PRONUNCIATION
nee-oh-DIM-ee-um

Overview

Neodymium was discovered in 1885 by Austrian chemist Carl Auer (Baron von Welsbach; 1858–1929). Auer found the new element in a mineral called didymia. Didymia, in turn, had been found in another complicated mineral known as ceria, originally found in Sweden in 1803. It took chemists nearly a century to completely analyze ceria. When they had done so, they found that it contained seven new elements. Neodymium was one of these.

Neodymium is in Row 6 of the periodic table. The periodic table is a chart that shows how chemical elements are related to each other. The elements in Row 6 are sometimes called the rare earth elements. The term "rare earth" is inaccurate, however. These elements are not especially rare but are difficult to separate from each other. The rare earth elements are also called the lanthanides. That name comes from the third element in Row 6, **lanthanum.**

Neodymium has long been used in coloring glass and is now used in making lasers, very powerful magnets, and special alloys.

Discovery and naming

During the late 1700s, two important mineral discoveries were made in Sweden. One was made just outside the town of Ytterby. The mineral found there, yttria, was eventually found to contain nine new elements. The second discovery was made near the town of Bastnas. That mineral, called cerite, was later found to contain seven new elements.

Cerite was thoroughly studied by Swedish chemist Carl Gustav Mosander (1797–1858). In 1839, Mosander was able to separate cerite into two parts, which he called **cerium** and lanthanum. Mosander believed he had found two new elements. Two years later, however, he learned that lanthanum was not an element but a mixture of two parts. Mosander called these two new parts lanthanum and didymium. Mosander chose the name didymium because it means "twin." He said that didymium was like an identical twin to lanthanum. Chemists later confirmed that two of Mosander's discoveries were really new elements: cerium and lanthanum.

Mosander's didymium was not an element, however. In 1885, Auer found that didymium consisted of two simpler materials. The new elements were named neodymium and **praseodymium.** Auer chose the name neodymium because it means "new twin." Praseodymium, by comparison, means "green twin." Compounds of praseodymium are green.

Mosander, Auer, and other chemists of the time had only crude equipment with which to work. They never isolated any new element in a pure form. They found compounds of the element, usually a compound of the element and **oxygen.** The first pure samples of neodymium were not produced until 1925.

Physical properties

Neodymium is a soft, malleable metal. Malleable means capable of being hammered into thin sheets. It can be cut and shaped fairly easily. It has a melting point of 1,024°C (1,875°F) and a boiling point of about 3,030°C (5,490°F). Neodymium has a density of 7.0 grams per cubic centimeter.

Chemical properties

Neodymium is somewhat reactive. For example, it combines with oxygen in the air to form a yellowish coating. To protect it from tarnishing, the metal is usually stored in mineral oil and wrapped in plastic.

Neodymium shows typical properties of an active metal. For example, it reacts with water and acids to release **hydrogen** gas.

WORDS TO KNOW

Isotopes two or more forms of an element that differ from each other according to their atomic weight

Lanthanides the elements in the periodic table with atomic numbers 58 through 71

Laser a device for producing very bright light of a single color

Malleable capable of being hammered into thin sheets

Periodic table a chart that shows how the chemical elements are related to each other

Reactive having a tendency to combine with other substances

Rare earth element *see* **Lanthanides**

Tarnishing oxidizing; reacting with oxygen in the air

Predicting volcanic eruptions using neodymium

Rare earth elements have very special applications in scientific research. For example, consider a discovery made by scientists at the Lawrence Berkeley Laboratory (LBL) Center for Isotope Geochemistry in Berkeley, California. These scientists were trying to predict the size of a volcanic eruption. If they knew an eruption was going to occur, could they estimate how much lava it would produce?

Surprisingly, they found the answer in isotopes of neodymium. They discovered that large volcanic eruptions produced lava with one kind of isotope composition. Smaller eruptions produced lava with a different isotope composition.

So when a volcano starts producing lava, it can be studied for neodymium isotopes. From the composition of isotopes, scientists may be able to predict how big the coming eruption will be. The LBL scientists hope to use this information to warn residents of the intensity of the eruption.

Occurrence in nature

Neodymium is one of the most abundant of the rare earth elements. Its abundance in the Earth's crust is thought to be about 12 to 24 parts per million. That places it about 27th among the chemical elements. It is slightly less abundant than **copper** and **zinc.**

The most common ores of neodymium are monazite and bastnasite. These ores are the most common source for all the rare earth elements.

Isotopes

Seven naturally occurring isotopes of neodymium are known. These isotopes are neodymium-142, neodymium-143, neodymium-144, neodymium-145, neodymium-146, neodymium-148, and neodymium-150. Six of these isotopes are stable and one, neodymium-144, is radioactive. Isotopes are two or more forms of an element. Isotopes differ from each other according to their mass number. The number written to the right of the element's name is the mass number. The mass number represents the number of protons plus neutrons in the nucleus of an atom of the element. The number of protons determines the element, but the number of neutrons in the atom of any one element can vary. Each variation is an isotope.

Neodymium combines with oxygen to form a yellowish coating. To protect it from tarnishing, the metal is usually stored in mineral oil and wrapped in plastic.

Neodymium-iron-boron (NIB) magnets are used in audio speakers.

Seven radioactive isotopes of neodymium are known also. A radioactive isotope is one that breaks apart and gives off some form of radiation. Radioactive isotopes are produced when very small particles are fired at atoms. These particles stick in the atoms and make them radioactive. None of neodymium's radioactive isotopes has any important use.

Extraction

Neodymium occurs with other rare earth elements in monazite, bastnasite, and allanite. It must first be separated from these

Neodymium is sometimes added to the glass in a lightbulb to filter out unwanted colors from the filament.

other elements. It is then obtained in a pure form by reacting neodymium fluoride (NdF_3) with calcium:

$$3NdF_3 + 2Ca \rightarrow 2CaF_2 + 3Nd$$

Uses and compounds

Neodymium and its compounds have a number of important uses. One is in a kind of laser known as a neodymium **yttrium aluminum** garnet (Nd:YAG) laser. A laser is a device for producing very bright and focused light of a single color. The Nd:YAG laser is used for treating bronchial cancer and certain eye disorders. The bronchi are air tubes that lead into the lungs.

Another important use of neodymium is in the manufacture of very strong magnets. The neodymium-**iron-boron** (NIB) magnet is one of the strongest magnets known. It is so strong it has to be handled with special care. Two NIB magnets can attract each other so strongly that they can smash into each other and shatter. An NIB magnet is inexpensive. An NIB magnet an inch in diameter and a quarter inch thick costs less than $10. Such magnets are used in stereo audio speakers.

Two neodymium-iron-boron (NIB) magnets can attract each other so strongly that they can smash into each other and shatter.

Neodymium is also used in various optical (light) devices. For example, the General Electric Company makes a lightbulb called an "Enrich" bulb. The glass of the bulb contains a small amount of neodymium that filters out yellowish and greenish colors from the filament. The filament is the metal wire that is heated and gives off light. The light produced by the Enrich bulb is a very bright white light.

One of the oldest uses of neodymium is in coloring glass. The addition of a small amount of the element to glass gives it a greenish color. Some Tiffany lamp shades contain neodymium.

Health effects

Neodymium is regarded as moderately hazardous. Its compounds are known to irritate the eyes and skin. It should be handled with caution.

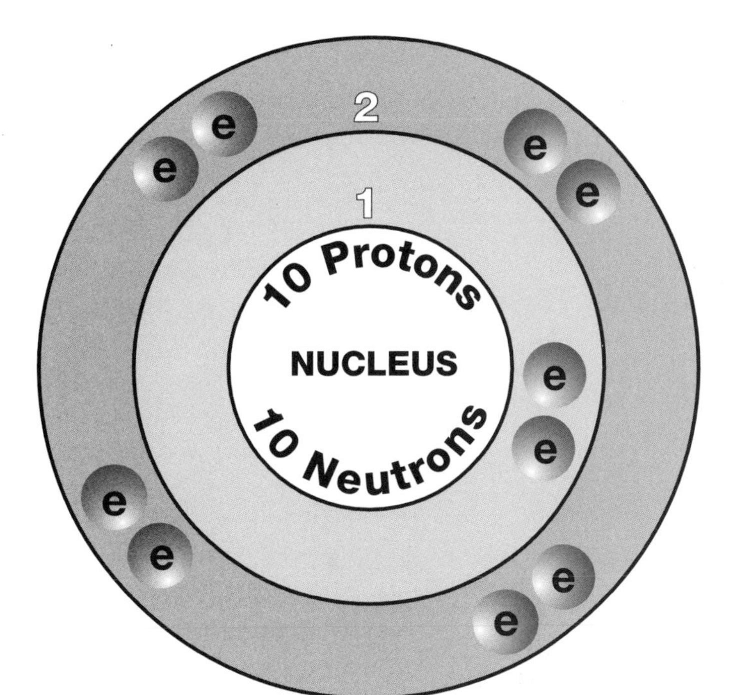

NEON

Overview

Neon is a member of the noble gas family. Other elements in this family includes **helium, argon, krypton, xenon,** and **radon.** These gases are in Group 18 (VIIIA) of the periodic table. The periodic table is a chart that shows how chemical elements are related to each other. The noble gases are sometimes called the inert gases. This name comes from the fact that these elements do not react very readily. In fact, compounds exist for only three noble gases—krypton, radon, and xenon. Chemists have yet to prepare compounds of helium, neon, or argon.

Neon was discovered in 1898 by British chemists William Ramsay (1852–1916) and Morris Travers (1872–1961). It occurs naturally in the atmosphere, but only in very small amounts.

Neon has relatively few uses. The most familiar is neon lighting. Today, neon signs of every color, shape, and size exist. Neon signs are often filled with neon gas, but they may also contain other gases as well. The gas contained in the sign tube determines the color of light given off. The color given off by neon itself is reddish-orange.

SYMBOL
Ne

ATOMIC NUMBER
10

ATOMIC MASS
20.179

FAMILY
Group 18 (VIIIA)
Noble gas

PRONUNCIATION
NEE-on

Discovery and naming

It took humans centuries to understand air. At one time, philosophers thought air was an element. Among the ancient Greeks, for example, the four basic elements were air, fire, water, and earth.

The first research to disprove that idea was done in the 1770s. In that decade, two new elements were discovered in air: **nitrogen** and **oxygen.** For some time, chemists were convinced that these two gases were the only ones present in air. That idea is easy to understand. Between them, nitrogen and oxygen make up more than 99 percent of air.

But over time, chemists became more skilled at making measurements. They recognized that something else was in air besides nitrogen and oxygen. That "something else" accounted for the remaining one percent that is not nitrogen or oxygen. In 1894, a third element was discovered in air: argon. Argon makes up about 0.934 percent of air. So, nitrogen, oxygen, and argon together make up about 99.966 percent of air.

But what was responsible for the remaining 0.034 percent of air? Chemists knew that other gases must be present in very small amounts. But what were those gases?

That question was answered between 1895 and 1900. Five more inert gases were discovered in air. One of those was neon.

Detecting gases in very small amounts was very difficult in the 1890s. Equipment was often not good enough to capture a tiny fraction of a milliliter of gas. But a new method, called spectroscopy, was developed that "sees" even small amounts of an element. Spectroscopy is the process of analyzing the light produced when an element is heated. The light pattern, or spectrum, produced is different for every element. The spectrum (plural: spectra) consists of a series of very specific colored lines.

In 1898, Ramsay and Travers were studying the minute amount of gas that remained after oxygen, nitrogen, and argon had been removed from air. They heated the sample of gas and studied the spectrum produced by it. Ramsay and Travers found spectral lines they had never seen before. They described their discovery:

WORDS TO KNOW

Isotopes two or more forms of an element that differ from each other according to their mass number

Laser a device for making very intense light of one very specific color that is intensified many times over

Noble gas an element in Group 18 (VIIIA) of the periodic table

Periodic table a chart that shows how the chemical elements are related to each other

Radioactive isotope an isotope that breaks apart and gives off some form of radiation

Spectroscopy the process of analyzing light produced when an element is heated

Spectra the lines produced when chemical elements are heated

A computer-generated model of a neon atom.

> The blaze of crimson light from the tube told its own story, and it was a sight to dwell upon and never to forget. It was worth the struggle of the previous two years; and all the difficulties yet to be overcome before the research was finished. The *undiscovered gas* had come to light in a manner which was no less than dramatic. For the moment, the actual spectrum of the gas did not matter in the least, for nothing in the world gave a glow such as we had seen.

Ramsay's son was one of the first people to hear about the discovery. He wanted to name the new element *novum,* meaning "new." His father liked the idea, but suggested using the Greek word for "new," *neos.* Thus, the element was named neon.

"The blaze of crimson light from the tube told its own story, and it was a sight to dwell upon and never to forget."

Physical properties

Neon is a colorless, odorless, tasteless gas. It changes from a gas to a liquid at −245.92°C (−410.66°F) and from a liquid to a solid at −248.6°C (−415.5°F). Its density is 0.89994 grams

A man bends a glass tube that will be used for neon lighting. The completed, glowing tubes are in the background.

per liter. By comparison, the density of air is about 1.29 grams per liter.

Chemical properties

Neon is chemically inactive. So far, it has been impossible to make neon react with any other element or compound.

Occurrence in nature and Extraction

The abundance of neon in normal air is 18.2 parts per million (0.0182 percent).

Isotopes

Three isotopes of neon exist, neon-20, neon-21, and neon-22. Isotopes are two or more forms of an element. Isotopes differ from each other according to their mass number. The number written to the right of the element's name is the mass number. The mass number represents the number of protons plus neutrons in the nucleus of an atom of the element. The number of protons determines the element, but the number of neutrons in the atom of any one element can vary. Each variation is an isotope.

Three radioactive isotopes of neon are known also. A radioactive isotope is one that breaks apart and gives off some form

The neon lights of Las Vegas, Nevada, in the early 1990s.

of radiation. Radioactive isotopes are produced when very small particles are fired at atoms. These particles stick in the atoms and make them radioactive.

None of the radioactive isotopes of neon has any commercial application.

Extraction

Neon can be obtained from air by fractional distillation. The first step in fractional distillation of air is to change a container of air to a liquid. The liquid air is then allowed to warm up. As the air warms, each element in air changes from a liquid back to a gas at a different temperature. The portion of air that changes back to a gas at −245.92°C is neon.

Uses

The best known use of neon gas is in neon lights. A neon light consists of a glass tube filled with neon or some other inert gas. An electric current is passed through the tube. The electric current causes neon atoms to break apart. After a fraction of a second, the parts recombine. When they recombine, they give off neon light. The light produced is the light given off by the neon light.

Neon lighting was invented by French chemist Georges Claude (1870–1960). Claude displayed his first neon sign at the Paris Exposition of 1910. He sold the first neon advertising sign to a Paris barber two years later.

By the 1920s, neon lighting had become popular in many parts of the world. Neon lights were fairly inexpensive, lasted a long time, and were very attractive. Probably the most spectacular collection of neon lighting is in Las Vegas, Nevada. Hotels, night clubs, and restaurants seem to try to outdo everyone else in having the biggest and brightest neon sign.

The first neon advertising sign was used by a Paris barber in 1912.

Neon lighting is now used for many other purposes. For example, neon tubes are used in instruments used to detect electric currents. Neon is also used in the manufacture of lasers. A laser is a device for producing very bright light of a single color. Lasers now have many uses in industry and medicine. They are very efficient at cutting metal and plastic. They can also be used to do very precise kinds of surgery.

Compounds

There are no compounds of neon.

Health effects

There are no known health effects of neon.

NEPTUNIUM

Overview

Neptunium lies in Row 7 of the periodic table. The periodic table is a chart that shows how chemical elements are related to one another. Neptunium is the first transuranium element. The term transuranium means "beyond **uranium.**" Any element with an atomic number greater than 92 (uranium's atomic number) is called a transuranium element. Elements in Row 7 are also called actinide elements. This name comes from the first element in Row 7, **actinium.**

Scientists have now found about 18 isotopes of neptunium. They are all radioactive. Neptunium was once a very rare element, but it can now be somewhat easily produced in a nuclear reactor. A nuclear reactor is a device in which nuclear fission reactions occur. Nuclear fission is the process of splitting atoms when neutrons collide with atoms of uranium or plutonium. These collisions produce new elements. Neptunium is used commercially only in specialized detection devices.

Discovery and naming

The discovery of neptunium in 1940 represented an important breakthrough in the study of chemical elements. Scientists had known for nearly a decade about an unusual kind of reaction.

SYMBOL
Np

ATOMIC NUMBER
93

ATOMIC MASS
237.0482

FAMILY
Actinide
Transuranium element

PRONUNCIATION
nep-TOO-nee-um

Neptunium is used in nuclear reactors. Pictured here is Three Mile Island in Middletown, Pennsylvania, site of a partial meltdown in 1979.

When an element is bombarded with neutrons, it sometimes changes into a new element. That new element has an atomic number one greater than the original element. For example, bombarding **copper** (atomic number 29) with neutrons may result in the production of **zinc** (atomic number 30). Bombarding **sodium** (atomic number 11) with neutrons may result in **magnesium** (atomic number 12).

One reason this discovery fascinated scientists was the possibility of bombarding uranium (atomic number 92) with neutrons. In the 1930s, uranium was the heaviest element known. It was the last element in the periodic table. But a "neutron change" like the ones described above would produce an element with atomic number 93. No one had ever heard of an element with atomic number 93!

In 1940, a pair of physicists at the University of California at Berkeley were studying this problem. Edwin M. McMillan (1907–91) and Philip H. Abelson (1913–) reported finding evi-

dence of element number 93. They suggested naming it neptunium, in honor of the planet Neptune. (Uranium, the element before neptunium, had been named for the planet Uranus.)

Physical and chemical properties

Neptunium is a silvery white metal with a melting point of 640°C (1,180°F) and a density of 20.45 grams per cubic centimeter.

Chemical properties

Neptunium is fairly reactive and forms some interesting compounds. Examples include neptunium dialuminide ($NpAl_2$) and neptunium beryllide ($NpBe_3$). These compounds are unusual because they consist of two metals joined to each other. Normally, two metals do not react with each other very easily. Neptunium also forms a number of more traditional compounds, such as neptunium dioxide (NpO_2), neptunium trifluoride (NpF_3), and neptunium nitrite ($NpNO_2$).

Occurrence in nature

When neptunium was first discovered, scientists thought it was an entirely artificial, or man-made, element. In 1942, very small amounts of the element were found in the Earth's crust. The element can sometimes be found in ores containing uranium and other radioactive elements.

Isotopes

All isotopes of neptunium are radioactive. Isotopes are two or more forms of an element. Isotopes differ from each other according to their mass number. The number written to the right of the element's name is the mass number. The mass number represents the number of protons plus neutrons in the nucleus of an atom of the element. The number of protons determines the element, but the number of neutrons in the atom of any one element can vary. Each variation is an isotope.

A radioactive isotope is one that breaks apart and gives off some form of radiation. Radioactive isotopes are produced when very small particles are fired at atoms. These particles stick in the atoms and make them radioactive.

The longest lived isotope is neptunium-237. Its half life is 2,140,000 years. The half life of a radioactive element is the time it takes for half of a sample of the element to break

WORDS TO KNOW

Actinide element *See* **Transuranium element**

Half life the time it takes for half of a sample of a radioactive element to break down

Isotopes two or more forms of an element that differ from each other according to their mass number

Nuclear fission the process that occurs when neutrons collide with atoms of uranium or plutonium, causing them to break apart

Nuclear reactor a device in which nuclear fission reactions occur

Radioactive isotope an isotope that breaks apart and gives off some form of radiation

Transuranium element an element with an atomic number greater than 92

One of the discoverers of neptunium, Edwin M. McMillan.

down. Of a sample of neptunium-237, only half would remain after 2,140,000 years. The other half would have broken down to form new elements.

Neptunium-239 is the only isotope of neptunium to have practical uses. It is used in special instruments for detecting the presence of neutrons.

Extraction

Pure neptunium metal can be made by heating neptunium trifluoride (NpF_3) with hot barium or lithium metal:

The case of the disappearing elements

Scientists think that the Earth was formed about five billion years ago. What elements would a chemist have found on the Earth in those days?

Part of the answer to that question is easy. Most of the elements found today were probably present five billion years ago. Those are the stable, or constant, elements. An untouched lump of gold in the Earth's crust five billion years ago would still be a lump of gold today.

But that statement is not true for radioactive elements. Radioactive elements "fall apart" spontaneously. They break down and form new, simpler elements.

The rate at which radioactive elements break down differs from element to element, however. Some break down slowly, others break down quickly. Scientists measure the rate of breakdown in half lives. An element with a long half life breaks down very slowly. An element with a short half life breaks down quickly.

Uranium, for example, has three naturally occurring isotopes. Their half lives are 4.6 billion years, 700 million years, and 25 million years. If 100 metric tons of uranium were present when the Earth was formed five billion years ago, about half of the first isotope would have broken down by now. About 50 metric tons of the element would remain. Scientists would have no trouble finding the element in the Earth's crust.

But neptunium is a different story. Its longest lived isotope is neptunium-237, with a half life of about two million years. If 100 million tonnes of neptunium were present at the Earth's beginning, only 50 million tons would be left after two million years. After another two million years (four million years altogether), only 25 million tons would be left. After another two million years (six million years altogether), only 12.5 million tons would be left.

Continue the mathematics. How much neptunium is left after 8 million years, 10 millions years, 12 million years, ... 5 *billion* years? No need to do the calculations: Not very much neptunium at all would be left! Perhaps, too little to even find in the Earth's crust.

So what does this example suggest about other transuranium elements, such as plutonium (number 94) and americium (number 95)? All of these elements have fairly short half lives. Of course, "fairly short" sometimes means "only" a few million years!

$$3Ba + 2NpF_3 \rightarrow 3BaF_2 + 2Np$$

The metal can now be purchased for legal uses from the Oak Ridge National Laboratory in Oak Ridge, Tennessee.

Uses and compounds

Neptunium and its compounds of neptunium have been made for research purposes. They are used in specialized detection devices and in nuclear reactors. Neither the element nor its compounds have any commercial uses.

Health effects

Neptunium is a very hazardous material. The radiation it gives off can cause serious health problems for humans and animals. It must be handled with great caution. Radiation transfers large amounts of energy to cells and is quite penetrating. Cells that are damaged, but not killed, often reproduce out of control. This growth by functionally damaged cells forms tumors and causes related problems for organs and tissues.

Neptunium is a very hazardous material. It must be handled with great caution.

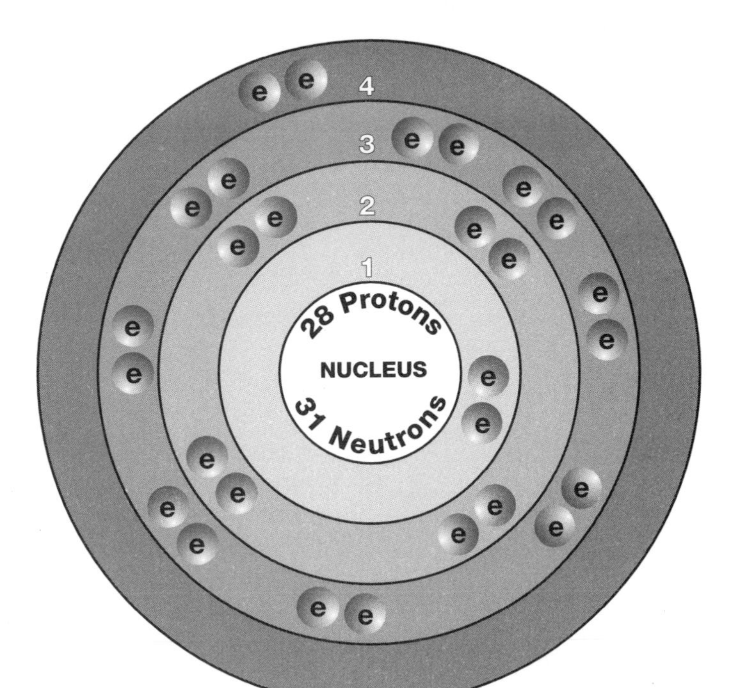

NICKEL

Overview

Nickel is the only element named after the devil. The name comes from the German word *Kupfernickel,* meaning "Old Nick's copper," a term used by German miners. They tried to remove copper from an ore that looked like copper ore, but they were unsuccessful. Instead of copper, they got slag, a useless mass of earthy material. The miners believed the devil ("Old Nick") was playing a trick on them. So they called the fake copper ore Old Nick's copper.

Since then, nickel has become a very valuable metal. The most common use is in the production of stainless steel, a strong material that does not rust easily. It is used in hundreds of industrial and consumer applications. Nickel is also used in the manufacture of many other alloys. An alloy is made by melting and mixing two or more metals. The mixture has properties different from those of the individual metals.

Nickel is classified as a transition metal. Transition metals are elements between Groups 2 (metals) and 13 (non-metals) in the periodic table. The periodic table is a chart that shows how chemical elements are related to one another. Nickel is closely

SYMBOL
Ni

ATOMIC NUMBER
28

ATOMIC MASS
58.69

FAMILY
Group 10 (VIIIB)
Transition metal

PRONUNCIATION
NI-kul

WORDS TO KNOW

Alloy a mixture of two or more metals with properties different from those of the individual metals

Ductile capable of being drawn into thin wires

Electroplating the process by which a thin layer of one metal is laid down on top of a second metal

Isotopes two or more forms of an element that differ from each other according to their mass number

Malleable capable of being hammered into thin sheets

Metallurgy the art and science of working with metals

Nickel allergy a health condition caused by exposure to nickel metal

Periodic table a chart that shows how the chemical elements are related to each other

Radioactive isotope an isotope that breaks apart and gives off some form of radiation

Toxic poisonous

Transition metal an element in Groups 3 through 12 of the periodic table

related to **iron, cobalt, copper,** and **zinc.** These metals are close to nickel in the periodic table.

Discovery and naming

The study of metals was difficult for early chemists. Many metals looked very similar. They also acted very much like each other chemically. Nickel was one of the metals about which there was much confusion.

Copper miners were confused about nickel and copper because they both occurred in ores with a green tint. But copper ores reacted differently to heat than did nickel ores. This confusion led to the choice for nickel's name.

But cobalt miners were confused too. Some ores of nickel also look like cobalt ores. But these ores did not react chemically in the same way either. Cobalt mine owners called the "misbehaving" ores of nickel "cobalt which had lost its soul."

Swedish mineralogist Axel Fredrik Cronstedt (1722–65) was the first person to realize that nickel was a new element. In 1751, he was given a new mineral from a cobalt mine near the town of Hälsingland, Sweden. While Cronstedt thought the ore might contain cobalt or copper, his tests produced a surprising result. He found something in the mineral that did not act like cobalt, copper, or any other known element. Cronstedt announced that he had found a new element. He used a shortened version of Kupfernickel for the name of the new element. He called it nickel.

Physical properties

Nickel is a silvery-white metal. It has the shiny surface common to most metals and is both ductile and malleable. Ductile means capable of being drawn into thin wires. Malleable means capable of being hammered into thin sheets. Its melting point is 1,555°C (2,831°F) and its boiling point is about 2,835°C (5,135°F). The density of nickel is 8.90 grams per cubic centimeter.

Nickel is only one of three naturally occurring elements that is strongly magnetic. The other two are iron and cobalt. But nickel is less magnetic than either iron or cobalt.

Chemical properties

Nickel is a relatively unreactive element. At room temperature, it does not combine with **oxygen** or water or dissolve in most

Nickel samples.

acids. At higher temperatures, it becomes more active. For example, nickel burns in oxygen to form nickel oxide (NiO):

$$2Ni + O_2 \text{ —heat→ } 2NiO$$

It also reacts with steam to give nickel oxide and hydrogen gas:

$$Ni + H_2O \text{ —heat→ } 2NiO + H_2$$

Occurrence in nature

Nickel makes up about 0.01 to 0.02 percent of the Earth's crust. It ranks about 22nd among the chemical elements in terms of abundance in the Earth's crust. Nickel is thought to be much more abundant in the Earth's core. In fact, many experts believe that the core consists almost entirely of iron and nickel.

One argument for this belief is the presence of nickel in meteorites. Meteorites are pieces of rock or metal from space that fall to the Earth's surface. Meteorites often contain a high percentage of nickel.

The most common ores of nickel include pentlandite, pyrrhotite, and garnierite. The element also occurs as an impurity in ores of iron, copper, cobalt, and other metals.

The United States' only nickel mine is located in Riddle, Oregon. In 1996, the mine produced 15,070 tons of nickel. By comparison, Russia produced 230,000 tons of nickel in the same year. Russia is the world's largest producer of nickel. Other major nickel producers are Canada (183,059 tons in 1996), New Caledonia (142,200 tons), Australia (113,134 tons), and Indonesia (90,000 tons).

The largest single deposit of nickel is located at Sudbury Basin, Ontario, Canada. The deposit was discovered in 1883. It covers an area 27 kilometers (17 miles) wide and 59 kilometers (37 miles) long. Some experts believe the deposit was created when a meteorite struck the earth at Sudbury Basin.

Isotopes

There are five naturally occurring isotopes of nickel: nickel-58, nickel-60, nickel-61, nickel-62, and nickel-64. Isotopes are two or more forms of an element. Isotopes differ from each other according to their mass number. The number written to the right of the element's name is the mass number. The mass number represents the number of protons plus neutrons in the nucleus of an atom of the element. The number of protons determines the element, but the number of neutrons in the atom of any one element can vary. Each variation is an isotope.

Seven radioactive isotopes of nickel are known also. A radioactive isotope is one that breaks apart and gives off some form of radiation. Radioactive isotopes are produced when very small particles are fired at atoms. These particles stick in the atoms and make them radioactive.

One radioactive isotope of nickel has limited use in industry, nickel-63. This isotope has two uses: for the detection of explosives, and in certain kinds of electronic devices, such as surge protectors. A surge protector is a device that protects sensitive electronic equipment like computers from sudden changes in the electric current flowing into them.

Extraction

The method used for making pure nickel metal is a common one in metallurgy. Metallurgy is the art and science of working with metals. Most nickel ores contain nickel sulfide (NiS). These ores are "roasted" (heated in air). Roasting converts the nickel sulfide to nickel oxide:

Ocean Song, *by John T. Scott. This sculpture, located in New Orleans, Louisiana, is an example of an artistic use of stainless steel.*

$$2NiS + 3O_2 \rightarrow 2NiO + 2SO_2$$

The nickel oxide is then treated with a chemical that will remove the oxygen from the nickel. For example:

$$2NiO + C \rightarrow CO_2 + 2Ni$$

A large amount of nickel is now recycled from scrap metal. Scrap metal comes from old cars, demolition of buildings, appliances like washing machines and stoves, and landfills. The task in recycling scrap metal is to find a way to separate

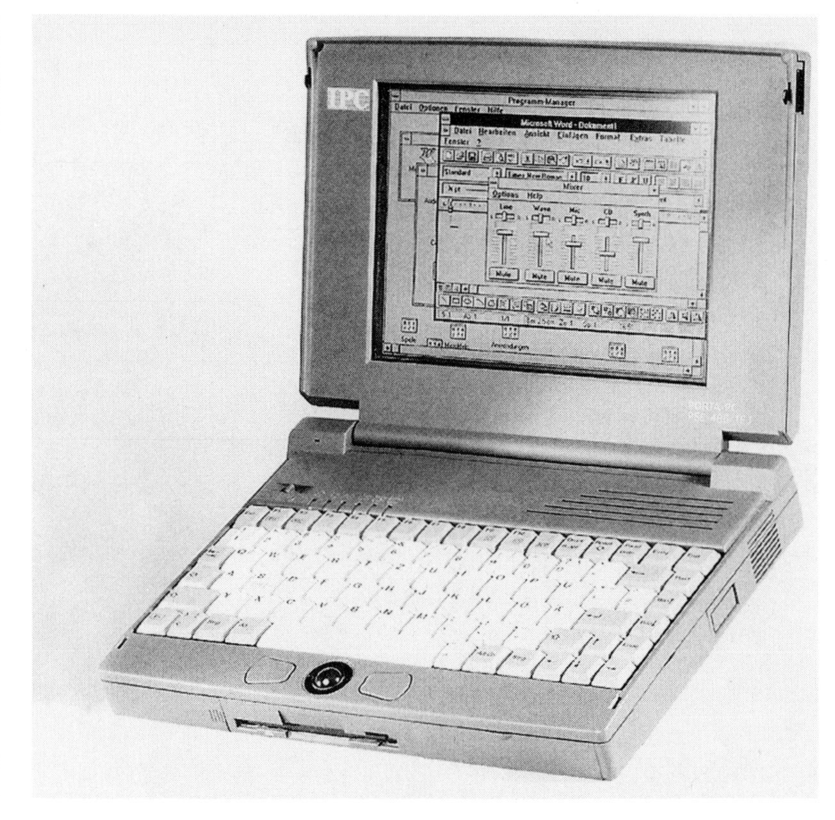

Nickel-cadmium batteries are used in laptop computers.

the nickel from other metals in the scrap. This can be done by taking advantage of special properties of nickel. For example, a magnet will remove nickel from scrap, leaving copper behind.

Uses

The most important use of nickel is in making alloys. About 80 percent of all nickel produced in the United States in 1996 was used to make alloys. About two-thirds of that amount went into stainless steel. Stainless steel is common to household appliances (like coffee makers, toasters, and pots and pans), kitchen sink tops and stoves, and medical equipment (X-ray machines, for example). It is also used to make heavy machinery and large containers in which large-scale chemical reactions are carried out. Artists sometimes use stainless steel in sculpture because it does not rust easily. Stainless steel is important to the food and beverage, petroleum, chemical, pharmaceutical (drug), pulp and paper, and textile industries.

Nickel is also used to make the superalloys used in jet engine parts and gas turbines. Superalloys are made primarily of iron,

Electroplating with nickel

Nickel is commonly used in electroplating. Electroplating is the process by which a thin layer of one metal is laid down on top of a second metal. Here is how electroplating is done.

First, the nickel compound to be laid down is dissolved in water. The solution may be nickel chloride ($NiCl_2$), nickel nitrate ($Ni(NO_3)_2$), or some other nickel compound.

Second, a sheet of the metal to be electroplated is placed into the solution. Suppose the metal is steel. The steel sheet is suspended in the nickel chloride, nickel nitrate, or other nickel solution.

Third, an electric current is passed through the solution. The current causes nickel to come out of the solution. The nickel is then deposited on the surface of the steel. The longer the current runs, the more nickel is laid down. The thickness of the nickel layer can be controlled by the time the electric current runs through the solution.

Electroplating is used to make metal products with very specific qualities. Steel is strong but tends to corrode easily. Nickel does not corrode as fast as steel. A thin layer of nickel on top of steel protects the steel from corrosion.

cobalt, or nickel. They also include small amounts of other metals, such as **chromium, tungsten, aluminum,** and **titanium.** Superalloys are resistant to corrosion (rusting) and retain their properties at high temperatures.

Nickel is also very popular in the manufacture of batteries. Nickel-cadmium (nicad) and nickel-metal hydride batteries are the most popular of these batteries. They are used in a great variety of appliances, including hand-held power tools, compact disc players, pocket recorders, camcorders, cordless and cellular telephones, scanner radios, and laptop computers.

Nickel is also used in electroplating, a process by which a thin layer of one metal is laid down on top of a second metal.

Compounds

Some nickel compounds have important uses also. Many of these compounds are used in electroplating. Some are used to make alloys of nickel. Other nickel compounds are used as coloring agents. For example, the compound nickel dimethylglyoxime ($C_8H_{14}N_4NiO_4$) is used as a coloring agent in paints, cosmetics, and certain kinds of plastics.

Other nickel compounds have somewhat more unusual uses. For example, the compound nickel dibutyldithiocarbamate ($Ni[CS_2N(C_4H_9)_2]_2$) is used as an antioxidant in tires. The rubber in tires reacts with oxygen in the air. When it does so, the rubber gets hard and stiff. The tires begin to break down. An additive like nickel dibutyldithiocarbamate can reduce the rate at which this process occurs. The life of tires is extended.

Health effects

Nickel can pose a health hazard to certain individuals. The most common health problem is called nickel allergy. Some people are more likely to develop nickel allergy than are others. People who are sensitive to nickel may develop a skin rash somewhat like poison ivy. The rash becomes itchy and may form watery blisters. Once a person gets nickel allergy, it remains with him or her forever.

Nickel is present in dozens of products. So it is easy for sensitive people to develop nickel allergy. Perhaps the most common cause of nickel allergy is body piercing. Some people have their ears pierced for earrings, while others have their lips, nose, or other body parts pierced. Inexpensive jewelry placed into these piercings is frequently made of stainless steel. Stainless steel contains nickel. The presence of nickel in a piercing can cause nickel allergy to develop.

Nickel can cause more serious health problems too. For example, people who are exposed to nickel fumes (dust and gas) breathe in nickel on a regular basis. Long term nickel exposure may cause serious health problems, including cancer.

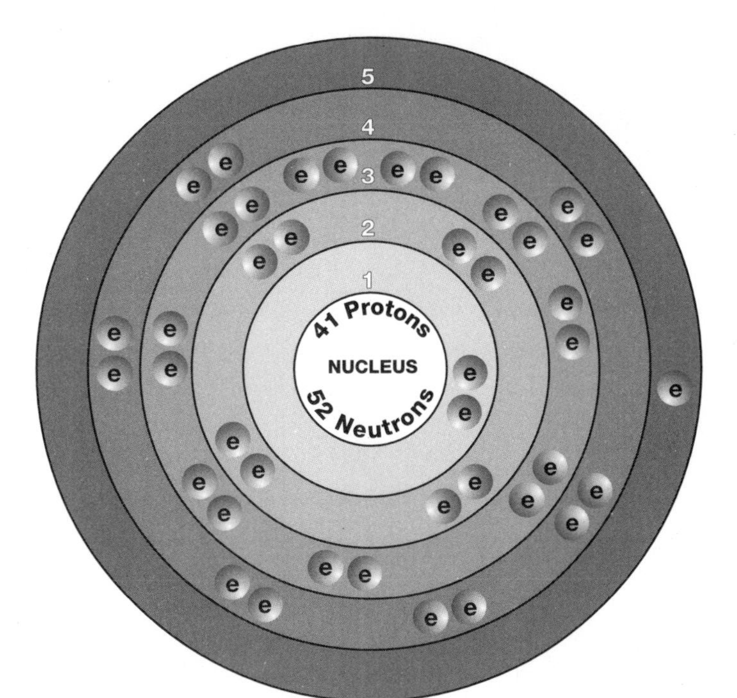

NIOBIUM

Overview

Niobium is a transition metal in Group 5 (VB) of the periodic table. The periodic table is a chart that shows how chemical elements relate to one another.

Niobium has a very interesting history. It was discovered by English chemist Charles Hatchett (1765–1847) in 1801. Hatchett found the element in a stone sent from North America. He named the element columbium. For years, scientists argued about the correct name for the element. Some still call the element columbium, although the official name is now niobium.

Niobium is used in many alloys. An alloy is made by melting and mixing two or more metals. The mixture has properties different from those of the individual metals. Niobium alloys are used in items that come into contact with the human body, such as rings for pierced ears, nose, and other body parts. Niobium is used in this kind of jewelry because it does not cause allergies or other problems.

SYMBOL
Nb

ATOMIC NUMBER
41

ATOMIC MASS
92.9064

FAMILY
Group 5 (VB)
Transition metal

PRONUNCIATION
nye-OH-bee-um

Discovery and naming

Historians today give credit for the discovery of niobium to Hatchett. The English chemist found the element in a "very heavy black

stone, with golden streaks" that he found in the British Museum. The stone had been sent to England from the United States by John Winthrop (1681–1747). Winthrop was a member of the British Royal Society, one of the most important scientific societies in the world. (He was also the grandson of the first governor of Connecticut and the great-grandson of the first governor of Massachusetts.) The rock had sat in the British Museum for nearly 70 years before anyone decided to analyze it. When Hatchett did so in 1801, he discovered a new element. He named it columbiun, after the mineral columbite in which it is often found.

Not everyone agreed with Hatchett's discovery at first. Some chemists were convinced that columbium was identical to the element **tantalum,** which had been discovered earlier. The confusion is easy to understand. The two elements have very similar properties and are difficult to separate. Finally, in 1844, German chemist Heinrich Rose (1795–1864) showed that tantalum and columbium really *were* different from each other. Rose then suggested the name niobium for the new element. The name comes from Greek mythology. Niobe is the daughter of the god Tantalus, from whom the name tantalum comes.

Scientists debated for nearly a century over which name to use. In 1949, niobium was officially adopted. However, many metallurgists (scientists who work with metals) still use the name columbium for the element.

Physical properties

Niobium is a shiny gray metal with a melting point of 2,468°C (4,474°F) and a boiling point of 4,927°C (8,901°F). Its density is 8.57 grams per cubic centimeter.

Chemical properties

Niobium metal is resistant to attack by most common chemicals. It does not combine with **oxygen** or most other active elements except at high temperatures. It does not react with most strong acids unless they are hot and concentrated.

Occurrence in nature

Niobium occurs primarily in two minerals, columbite and pyrochlore. The original name columbium was taken from the first of these minerals. Niobium always occurs with tantalum in these minerals. Separating the two elements is always the most difficult step in their preparation.

WORDS TO KNOW

Alloy a mixture of two or more metals that has properties different from those of the individual metals

Columbium an alternative name for niobium

Hypoallergenic not causing an allergic reaction

Isotopes two or more forms of an element that differ from each other according to their mass number

Radioactive isotope an isotope that breaks apart and gives off some form of radiation

Superconductivity the tendency of an electric current to flow through a material without resistance

Niobium samples.

Columbite and pyrochlore are not mined in the United States. These two minerals are imported primarily from Brazil and Canada.

Scientists believe that niobium's abundance in the Earth's crust is about 20 parts per million. That makes it about as abundant as **nitrogen** and **lithium,** and slightly more abundant than **lead.**

Isotopes

Only one naturally occurring isotope of niobium exists, niobium-93. Isotopes are two or more forms of an element. Isotopes differ from each other according to their mass number. The number written to the right of the element's name is the mass number. The mass number represents the number of protons plus neutrons in the nucleus of an atom of the element. The number of protons determines the element, but the number of neutrons in the atom of any one element can vary. Each variation is an isotope.

Skateboards often include niobium steel components.

At least a dozen radioactive isotopes of niobium are known also. A radioactive isotope is one that breaks apart and gives off some form of radiation. Radioactive isotopes are produced when very small particles are fired at atoms. These particles stick in the atoms and make them radioactive.

None of the radioactive isotopes of niobium have any practical application.

Extraction

The first step in preparing niobium metal is to separate its

Niobium alloys are used in superconducting magnets. Here, a small magnet levitates over a cooled slab of superconducting ceramic.

compounds from tantalum in an ore. The niobium compounds are then heated in air to change them to niobium oxide (Nb_2O_5). This is then heated with charcoal **(carbon)** to produce free metal:

$$2Nb_2O_5 + 5C \rightarrow 5CO_2 + 4Nb$$

Uses

Niobium is used primarily in making alloys. For example, the addition of niobium to steel greatly increases its strength. One use of such steel is in the construction of nuclear reactors.

Nuclear reactors are devices in which the energy of nuclear reactions is converted to electricity. Niobium steel is used because it keeps its strength at the very high temperatures produced there.

The demand for niobium steel has increased. One reason is its increased use in airplanes and space vehicles. Some skateboards also include niobium steel components.

Another popular use of niobium alloys is in the making of jewelry. These alloys are lightweight and hypoallergenic. The term hypoallergenic means that they do not cause skin reactions. People who wear pierced earrings or similar forms of jewelry will not develop skin problems from niobium alloy jewelry.

Niobium alloys are used in jewelry because they are lightweight and hypoallergenic (won't cause skin reactions).

Niobium alloys are also used in the construction of superconducting magnets. A superconducting material is one that has no resistance to an electric current. Once an electric current begins to flow in such a material, it continues to flow practically forever.

The most powerful magnets in the world are those made with superconducting materials. In 1997, a new record was set for superconducting magnets at the Lawrence Berkeley Laboratory in Berkeley, California. A magnet made with an alloy of niobium and **tin** proved to be three times as strong as the best magnet previously known.

Compounds

Niobium diselenide ($NbSe_2$) is sometimes used as a lubricant at high temperatures. It does not break down at temperatures up to about 1300°C. Niobium silicide ($NbSi_2$) is used as a refractory material. A refractory material is one that can withstand very high temperatures.

Health effects

Neither niobium nor its compounds are known to pose serious health effects for humans and animals.

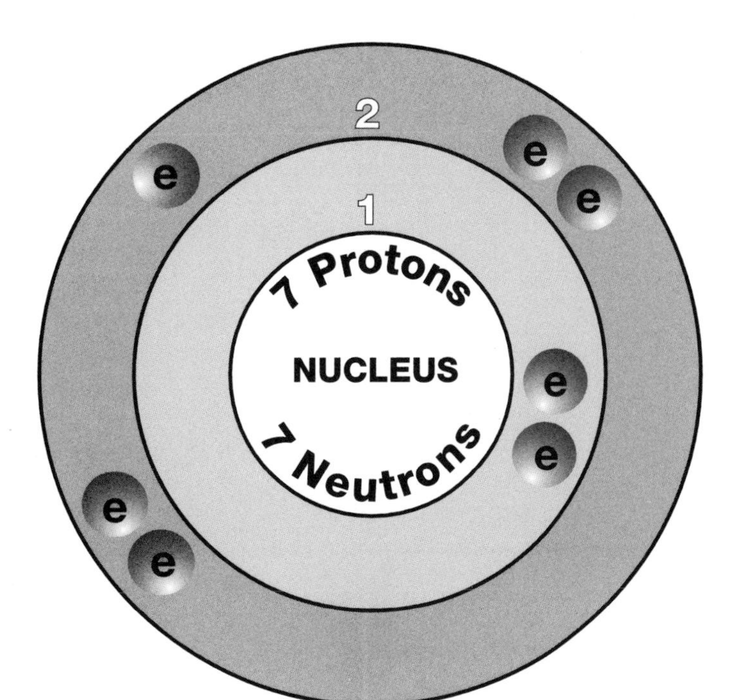

NITROGEN

Overview

Nitrogen is the first member in Group 15 (VA) of the periodic table. The periodic table is a chart that shows how chemical elements are related to one another. Nitrogen is in a family group named after itself. Other elements in the nitrogen family are **phosphorus, arsenic, antimony,** and **bismuth.**

Nitrogen is one of the most interesting of all chemical elements. It is not a very active element. It combines with relatively few other elements at room temperature. Yet, the compounds of nitrogen are enormously important both in living organisms and in industrial applications. Five of the top fifteen chemicals that are produced synthetically by chemical producers are compounds of nitrogen or the element itself. How does such an inactive element end up with so many important compounds?

Nitrogen makes up more than three-quarters of the Earth's atmosphere. It is also found in a number of rocks and minerals in the Earth's surface. It ranks about number 32 among the elements in terms of abundance in the Earth's crust.

Nitrogen was discovered by a number of chemists at about the same time, approximately 1772. But it was not until the early

SYMBOL
N

ATOMIC NUMBER
7

ATOMIC MASS
14.0067

FAMILY
Group 15 (VA)
Nitrogen

PRONUNCIATION
NYE-tru-jun

part of the twentieth century, when chemists learned how to make compounds of nitrogen, that the most important uses of the element became known.

By far the most notable use of nitrogen is in the production of ammonia (NH_3). Ammonia is used to make other compounds, such as ammonium sulfate ($(NH_4)_2SO_4$), ammonium nitrate (NH_4NO_3), urea ($CO(NH_2)_2$), and nitric acid (HNO_3). These compounds are primarily used to make synthetic fertilizer. Both elemental nitrogen and nitrogen compounds have a number of important industrial uses.

Discovery and naming

Gases were poorly understood by chemists until the late 1700s. What is air "made of?" That question is difficult to answer for a number of reasons. First, air cannot really be "seen." In fact, it took chemists many years to figure out how to capture air so that they could study it. Also, is ordinary "air" an element or a compound? For many centuries, philosophers said that air was an element. They could not imagine how anything as basic as air could be made of other materials.

Also, is ordinary air different from other kinds of "airs" seen in nature? For example, "air" sometimes comes bubbling out of the ground near oil wells. Today, scientists know that kind of "air" as methane gas (CH_4). But early chemists were not sure how "oil air" differed from ordinary air.

Some important breakthroughs in the study of air occurred in the 1770s. The key was a simple experiment that science students still do today. The experiment begins with an empty bottle being turned upside down in a pan of water. The air in the bottle cannot get out.

If a burning candle is placed inside the bottle with the trapped air, the water rises just a bit. Why does this happen? Early chemists thought that a part of the air was used up as the candle burns. Today, they know that part of the air is **oxygen** gas. Another part of the air is left behind. That part does not disappear when the candle burns.

This simple experiment shows that air is composed of (at least) two different elements: oxygen and something else. One of the first people to discover what the "something else" is

WORDS TO KNOW

Anhydrous ammonia dry ammonia gas

Catalyst a substance used to speed up or slow down a chemical reaction without undergoing any change itself

Isotopes two or more forms of an element that differ from each other according to their mass number

Nitrogen fixation the process of converting nitrogen as an element to a compound that contains nitrogen

Periodic table a chart that shows how the chemical elements are related to each other

Proteins compounds that are vital to the building and growth of cells

Radioactive isotope an isotope that breaks apart and gives off some form of radiation

Tracer a radioactive isotope whose presence in a system can easily be detected

was Scottish physician and chemist Daniel Rutherford (1749–1819). Rutherford carried out an experiment like the candle-in-a-bottle research just described.

Some of the greatest chemists of the time were working on this problem at the same time that Rutherford made his discovery. English chemist Henry Cavendish (1731–1810) probably discovered nitrogen before Rutherford did, but did not publish his findings. And in science, the first person to publish the results of an experiment usually gets credit for the work.

It seems likely that English chemist Joseph Priestley (1733–1804) and Swedish chemist Carl Wilhelm Scheele (1742–86) also discovered nitrogen in the early 1770s. (See sidebar on Scheele in the **chlorine** entry in Volume 1.)

Chemists debated about the name of this new element for some time. Antoine-Laurent Lavoisier (1743–94), a French chemist and the "father of modern chemistry," preferred the name azote meaning "without life." Lavoisier chose this name because nitrogen does not support breathing, the way oxygen does. (See sidebar on Lavoisier in the **oxygen** entry in Volume 2.)

The modern name of nitrogen was first suggested in 1790 by French chemist Jean Antoine Claude Chaptal (1756–1832). This name made sense to chemists when they realized that the new gas was present in both nitric acid and nitrates. Thus, nitrogen means "nitrate and nitric acid" *(nitro-)* and "origin of" *(-gen)*.

Physical properties

Nitrogen is a colorless, odorless, tasteless gas with a density of 1.25046 grams per liter. By comparison, the density of air is about 1.29 grams per liter. Nitrogen changes from a gas into a liquid at a temperature of −195.79°C (−320.42°F). It changes from a liquid to a solid at a temperature of −210.01°C (−346.02°F). When it freezes, it becomes a white solid that looks like snow. Nitrogen is slightly soluble in water. About two liters of nitrogen can be dissolved in 100 liters of water.

Chemical properties

At room temperature, nitrogen is a very inactive gas. It does not combine with oxygen, **hydrogen,** or most other elements. Nitrogen will combine with oxygen, however, in the presence of lightning or a spark. The electrical energy from either of those sources causes nitrogen and oxygen to form nitric oxide:

$$N_2 + O_2 \text{ —electrical energy} \rightarrow 2NO$$

Nitric oxide is more active than free nitrogen. For example, nitric oxide combines with oxygen and water in the atmosphere to make nitric acid. When it rains, nitric acid is carried to the earth. There it combines with metals in the Earth's crust. Compounds known as nitrates and nitrites are formed.

Changing nitrogen as an element to nitrogen in compounds is called nitrogen fixation. The reaction between nitrogen and oxygen in the air when lightning strikes is an example of nitrogen fixation.

Nitrogen combines with oxygen in the presence of lightning or a spark. The electrical energy from those sources causes nitrogen and oxygen to form nitric oxide.

Certain bacteria have developed methods for fixing nitrogen. These bacteria live on the root hairs of plants. They take nitrogen out of air dissolved in the ground and convert it to compounds, such as nitrates. Those nitrates are used to make protein molecules, compounds vital to the building and growth of cells.

Plants, animals, and humans do not have the ability to fix nitrogen. All living organisms on Earth depend on soil bacteria to carry out this process. Plants can grow because the bacteria fix nitrogen for them. They use the fixed nitrogen to make proteins. Animals and humans can survive because they eat plants. They also depend on the soil bacteria that allow plants to make proteins. So all living creatures rely on soil bacteria to fix their nitrogen for them and, therefore, to survive.

Occurrence in nature

Nitrogen is a fairly common element in the Earth's crust. It occurs primarily as nitrates and nitrites. Nitrogen is by far the most important element in the Earth's atmosphere. It makes up 78.084 percent of the atmosphere.

Isotopes

Two naturally occurring isotopes of nitrogen exist, nitrogen-14 and nitrogen-15. Isotopes are two or more forms of an element. Isotopes differ from each other according to their mass number. The number written to the right of the element's name is the mass number. The mass number represents the number of protons plus neutrons in the nucleus of an atom of the ele-

ment. The number of protons determines the element, but the number of neutrons in the atom of any one element can vary. Each variation is an isotope.

Five radioactive isotopes of nitrogen are known also. A radioactive isotope is one that breaks apart and gives off some form of radiation. Radioactive isotopes are produced when very small particles are fired at atoms. These particles stick in the atoms and make them radioactive.

None of the radioactive isotopes of nitrogen has any important commercial use. However, nitrogen-15 is used quite often in tracer studies. A tracer is a radioactive isotope whose presence in a system can be detected. Normally, tracer studies use radioactive isotopes. These isotopes give off radiation that can be detected with instruments. Nitrogen-15 is used for a different reason. A compound made with nitrogen-15 will weigh just a little bit more than one made with nitrogen-14. There are simple chemical methods for detecting whether a heavier compound or a lighter one is present in a system. Thus, nitrogen-15 can be used to trace the path of nitrogen through a system.

Extraction

Nitrogen is almost always made from liquid air. Liquid air is made by cooling normal atmospheric air to very low temperatures. As the temperature drops, the gases contained in air turn into liquids. At −182.96°C (−297.33°F), oxygen changes from a gas into a liquid. At −195.79°C (−320.42°F), nitrogen changes from a gas into a liquid. And so on. Eventually, all the gases in air can be made to liquefy (change into a liquid).

The reverse process also takes place. Suppose liquid air in a container warms up slowly. When its temperature reaches −195.79°C, liquid nitrogen changes back to a gas. A container can be put into place to catch the nitrogen as it boils off the liquid air. When the temperature reaches −182.96°C, oxygen changes from a liquid back to a gas. Another container can be put into place. The escaping oxygen can be collected. All of the gases in atmospheric air can be produced by this method.

Large amounts of nitrogen gas are produced in this way. In fact, nitrogen is second only to sulfuric acid in terms of production. In 1996, more than a trillion cubic feet of nitrogen gas were produced in the United States alone.

Uses

Nitrogen gas is used where an inert atmosphere is needed. An inert atmosphere is one that does not contain active elements. Ordinary air is not an inert atmosphere. It contains oxygen. Oxygen tends to react with other elements.

Suppose an ordinary light bulb were filled with air. When the bulb is turned on, an electric current runs through the metal filament (wire) inside the bulb. The filament gets very hot, begins to glow, and gives off light.

But a hot metal wire will react quickly with oxygen in ordinary air. The metal combines with oxygen to form a compound of the metal. The metal compound will not conduct an electric current. The bulb will "burn out" very quickly.

Historic documents like the Declaration of Independence are protected in air-tight containers filled with nitrogen gas. This keeps them from oxygen and other gases that would cause them to decay.

An easy solution to that problem is to use nitrogen instead of ordinary air in the light bulb. Nitrogen does not react with other elements very well, even when they get hot. The filament can get very hot, but the metal of which it is made will not combine with nitrogen gas. The nitrogen gas is an inert atmosphere for the bulb.

Another use for inert atmospheres is in protecting historic documents. Suppose the Declaration of Independence were simply left on top of a table for people to see. The paper and ink in the document would soon begin to react with oxygen in the air. They would both begin to decay. Before long, the document would begin to fall apart.

Instead, important documents like the Declaration of Independence are kept in air-tight containers filled with nitrogen gas. The documents are protected from oxygen and other gases in the air with which they might react.

Fairly simple methods are now available for changing nitrogen gas into liquid nitrogen. Liquid nitrogen is used to freeze other materials. The temperature of the nitrogen has to be reduced to −195.79°C (−320.42°F) for this change to occur.

Today, it is possible to buy large containers of liquid nitrogen. The liquid nitrogen can be used, then, to freeze other materials. For example, foods can be frozen simply by dipping them into large vats of liquid nitrogen. The frozen foods in a grocery store are usually produced this way. Liquid nitrogen can also

Liquid nitrogen is used to freeze foods. Frozen foods commonly found in grocery stores are produced this way.

be used to keep foods cold when they are being transported from one place to another.

Compounds

Nitrogen is the starting point for an important group of compounds. First, nitrogen is combined with hydrogen to make ammonia (NH_3). The production of ammonia is sometimes called industrial nitrogen fixation.

The formation of ammonia from nitrogen and hydrogen is very difficult to accomplish. The two elements do not easily combine. Finding a way to make nitrogen and hydrogen combine was one of the great scientific discoveries of the twentieth century.

The nitrogen compound ammonia is used by most farmers in synthetic fertilizers to ensure large crops.

That discovery was made by German chemist Fritz Haber (1868–1934) in 1905. He found that nitrogen and hydrogen would combine if they were heated to a very high temperature with a very high pressure. He also found that a catalyst was needed to make the reaction occur. The catalyst he used was **iron** metal, though other metals are sometimes used. A catalyst is a substance used to speed up or slow down a chemical reaction. The catalyst does not undergo any change itself during the reaction.

A farmer sprays nitrogen fertilizer on his rice field in California.

The significance of Haber's discovery was soon apparent. World War I began in 1914. Very soon, Germany was no longer able to get nitrates from Chile. Nitrates were essential to the war effort for making explosives. But ships carrying nitrates from Chile to Germany were usually not able to get across the Atlantic Ocean.

However, nitrates can be made from ammonia using the Haber process. Germany was then able to make all the ammonia it needed. The ammonia was converted to nitrates for explosives. So a lack of nitrates from Chile did not stop the German war machine. Instead, it went on fighting for another four years.

Ammonia usually ranks about number 5 or 6 among the most highly produced chemicals in the United States. The most important use of ammonia is in synthetic fertilizers. A synthetic fertilizer is a mixture of compounds used to make plants grow better. Most farmers use huge amounts of synthetic fertilizer every year to ensure large crops.

Nitrogen's role in the Oklahoma City bombing

On April 19, 1995, a bomb exploded at the Alfred P. Murrah Federal Building in Oklahoma City, Oklahoma. The devastating impact of the explosion destroyed the building within eight seconds. Each of the nine floors collapsed on top of one another. The bomb killed 168 people and injured hundreds more. The nation was stunned to learn that the attack had not been committed by international terrorists, but by an American, Timothy McVeigh.

While searching the home of McVeigh's accomplice (a partner in a crime), Terry Nichols, investigators found a receipt for ammonium nitrate. This nitrogen compound is most commonly purchased by farmers. They use it as a fertilizer for their crops. But it can also be used as an explosive. The amount on the receipt was for 2,000 pounds. Since neither Nichols nor McVeigh was farming at the time the fertilizer was purchased, there was no real reason to purchase such a large amount of fertilizer.

Investigators determined that an ammonium nitrate bomb did, indeed, destroy the Murrah Building. The explosion resulted from a mixture of 4,800 pounds of ammonium nitrate and fuel oil from twenty plastic drums.

In June 1997, McVeigh was found guilty of all eleven charges against him, including eight counts of first-degree murder of federal agents. He was sentenced to death later that month.

In 1996, more than 38 million tons of nitrogen-containing synthetic fertilizer was made in the United States. Nearly half of that was anhydrous ammonia. Anhydrous means "without water." Anhydrous ammonia is simply ammonia gas. It is stored in large tanks. Farmers inject anhydrous ammonia directly into the ground to produce strong and healthy plants.

Ammonia is also found in many household cleaners, especially glass-cleaning and grease-cutting products.

The largest producer of ammonia in the world is China. Other large producers are the United States, India, Russia, Canada, Ukraine, Indonesia, and the Netherlands.

Ammonia can also be converted into other forms. For example, it can be combined with nitric acid (HNO_3) to form ammonium nitrate (NH_4NO_3). And it can be combined with sulfuric acid (H_2SO_4) to make ammonium sulfate ($(NH_4)_2SO_4$):

$$NH_3 + HNO_3 \rightarrow NH_4NO_3$$

$$2NH_3 + H_2SO_4 \rightarrow (NH_4)_2SO_4$$

In 1996, about 2.9 million tons of ammonium nitrate and 2.6 million tons of ammonium sulfate were produced as fertilizers. These two compounds usually rank about number 14 and number 30 among chemicals produced in the United States

Ammonium nitrate and ammonium sulfate both have other uses also. For example, ammonium nitrate is used to make explosives, fireworks, insecticides and herbicides (chemicals that kill insects and weeds), and rocket fuel. Ammonium sulfate is also used in water treatment systems, as a food additive, in the tanning of leather, in fireproofing materials, and as a food additive.

Nitrogen is absolutely essential to all living organisms. It is an important part of all protein molecules.

Yet another important compound of nitrogen is nitric acid (HNO_3). Nitric acid is made by reacting ammonia with oxygen:

$$NH_3 + 2O_2 \xrightarrow{heat} HNO_3 + H_2O$$

Nitric acid usually ranks about number 13 among chemicals produced in the United States. The major use of nitric acid is to make ammonium nitrate as a synthetic fertilizer. Nitric acid is also used to make explosives, dyes, certain kinds of synthetic rubber and plastics, and in the preparation of metals.

Health effects

Nitrogen is absolutely essential to all living organisms. It is an important part of all protein molecules. Proteins are the building material in all kinds of cells. They are also used for many other functions. For example, all living organisms use hormones to send chemical messages from one cell to another. Hormones are proteins.

Nitrogen is also used to make nucleic acids. Nucleic acids have many important functions in living organiams. For one thing, they store the organism's genetic information. The genetic information is the set of instructions that tell every cell what its job in the organism is. It passes on that information from one generation to the next.

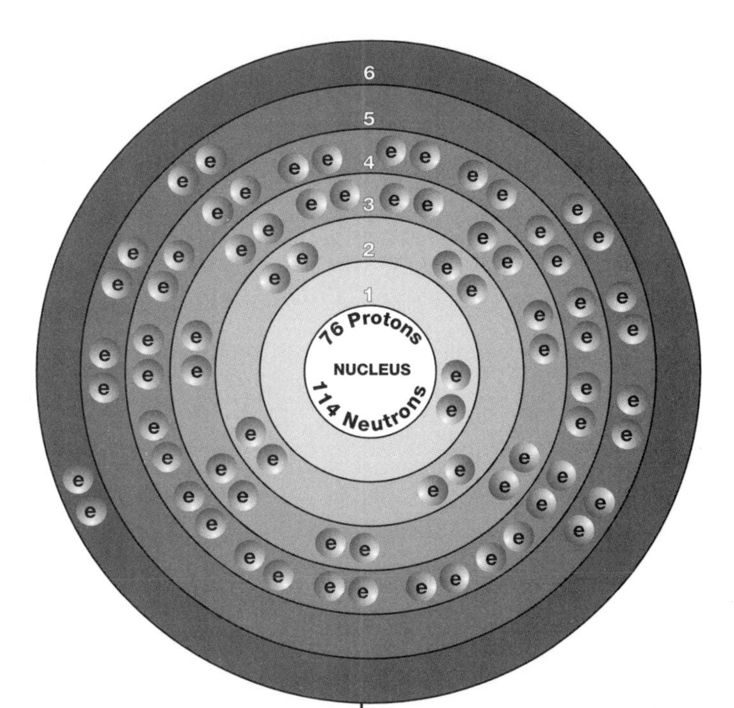

OSMIUM

SYMBOL
Os

ATOMIC NUMBER
76

ATOMIC MASS
190.2

FAMILY
Group 8 (VIIIB)
Transition metal
Platinum group

PRONUNCIATION
OZ-mee-um

Overview

Osmium is an element in Group 8 (VIIIB) of the periodic table. The periodic table is a chart showing how chemical elements are related to one another. Osmium is also a member of the platinum family. This family consists of five other elements: **ruthenium, rhodium, palladium, iridium,** and **platinum.** These elements often occur together in the Earth's crust. They also have similar physical and chemical properties, and are used in alloys.

Osmium was discovered in 1804 by English chemist Smithson Tennant (1761–1815). Tennant found the new element in an ore of platinum.

Osmium is a very rare element and has few commercial uses. Osmium tetroxide (OsO_4), is more widely used, however, because it is so active.

Discovery and naming

Platinum metal (atomic number 78) was known to chemists as early as 1741. Over the next 60 years, however, scientists discovered that the substance they knew as "platinum" was usually a mixture of substances. These substances proved to be new elements. Osmium was one of the new elements discovered in impure platinum.

In the early 1800s, Smithson Tennant was studying platinum. He found that a black powder remained when platinum was dissolved in aqua regia. Aqua regia is a mixture of hydrochloric and nitric acids. The term aqua regia means "royal water." It often dissolves materials that either acid by itself does not dissolve.

In 1804, Tennant announced that the black powder was actually a mixture of two new elements. He called them iridium and osmium. He suggested osmium's name because of the unusual smell of the compound he was working with, osmium tetroxide. Osmium comes from the Greek word *osme,* meaning "odor."

Physical properties

Osmium is a bluish-white, shiny metal with a melting point of about 3,000°C (5,400°F) and a boiling point of about 5,500°C (9,900°F). Its density is 22.5 grams per cubic centimeter. These numbers are the highest of any platinum metal. They are also among the highest of all elements.

Osmium is unworkable as a metal. It cannot be melted and shaped like most metals. Because it is unworkable, it has very few practical uses.

Chemical properties

Osmium is dissolved by acids or by aqua regia only after long periods of exposure to the liquids. When heated, the metal combines with **oxygen** to form osmium tetroxide (OsO_4). Osmium tetroxide is very toxic and the only important commercial compound of osmium.

Occurrence in nature

Osmium is very rare. Its abundance is thought to be about 0.001 parts per million (one part per billion). That places the element among the half dozen least abundant elements in the Earth's crust.

The most common ore of osmium is osmiridium. The element also occurs in all ores of platinum.

Isotopes

There are seven naturally occurring isotopes of osmium. The most abundant are osmium-192, osmium-190, and osmium-

WORDS TO KNOW

Alloy an alloy is a mixture of two or more metals that has properties different from those of the individual metals

Aqua regia a mixture of hydrochloric and nitric acids that dissolves many materials that are not dissolved by either acid by itself

Catalyst a substance used to speed up or slow down a chemical reaction without undergoing any change itself

Isotopes two or more forms of an element that differ from each other according to their mass number

Radioactivity having a tendency to break apart and give off some form of radiation

Toxic dangerous

Osmium is often added to platinum to create an alloy used in pen points.

189. These three isotopes make up 41.0 percent, 26.4 percent, and 16.1 percent of natural osmium, respectively. Isotopes are two or more forms of an element. Isotopes differ from each other according to their mass number. The number written to the right of the element's name is the mass number. The mass number represents the number of protons plus neutrons in the nucleus of an atom of the element. The number of protons determines the element, but the number of neutrons in the atom of any one element can vary. Each variation is an isotope.

Many radioactive isotopes of osmium are known also. A radioactive isotope is one that breaks apart and gives off some form of radiation. Radioactive isotopes are produced when very small particles are fired at atoms. These particles stick in the atoms and make them radioactive. No radioactive isotope of osmium has any important use.

Extraction

Osmium is obtained when platinum metal is extracted from its ores.

Uses

Osmium metal has few uses. It is sometimes added to platinum or iridium to make them harder. Some of the best pen points,

for example, are made of osmium-platinum alloys. An alloy is made by melting or mixing two or more metals. The osmium-platinum alloy is harder than pure platinum. Some alloys of osmium and platinum are also used to make specialized laboratory equipment.

Finely divided osmium metal is also used as a catalyst. A catalyst is a substance used to speed up or slow down a chemical reaction. The catalyst does not undergo any change itself during the reaction. The process for making ammonia from combined **hydrogen** and **nitrogen** sometimes uses osmium as a catalyst.

Osmium tetroxide is so dangerous to use that it is shipped in a small glass container. The container carries no label or ink because each would react violently with the compound.

Compounds

Osmium tetroxide (OsO_4) is in demand for use as a catalyst for research purposes. The problem is that this compound of osmium is very dangerous to use. It is shipped in small glass containers called ampules. The ampules carry no labels, nor are they marked with ink. The label and ink would react violently with osmium tetroxide. Users are instructed to open and use an ampule containing osmium tetroxide with great care.

Health effects

Some compounds of osmium are extremely dangerous. They irritate the respiratory passage (throat, lungs, etc.), the skin, and the eyes. They must be handled with extreme care. This caution is especially important for the most widely used compound of osmium, osmium tetroxide.

OXYGEN

Overview

Oxygen is the first element in Group 16 (VIA) of the periodic table. The periodic table is a chart that shows how chemical elements are related to each other. The elements in Group 16 are said to belong to the chalcogen family. Other elements in this group include **sulfur, selenium, tellurium,** and **polonium.** The name chalcogen comes from the Greek word *chalkos,* meaning "ore." The first two members of the family, oxygen and sulfur, are found in most ores.

Oxygen is by far the most abundant element in the Earth's crust. Nearly half of all the atoms in the earth are oxygen atoms. Oxygen also makes up about one-fifth of the Earth's atmosphere. Nearly 90 percent of the weight of the oceans is due to oxygen. In addition, oxygen is thought to be the third most abundant element in the universe and in the solar system.

The discovery of oxygen is usually credited to Swedish chemist Carl Wilhelm Scheele (1742–86) and English chemist Joseph Priestley (1733–1804). The two discovered oxygen at nearly the same time in 1774, working independently of each other.

Oxygen is necessary for the survival of all animal life on Earth. Animals breathe in oxygen and breathe out carbon dioxide.

SYMBOL
O

ATOMIC NUMBER
8

ATOMIC MASS
15.9994

FAMILY
Group 16 (VIA)
Chalcogen

PRONUNCIATION
OK-si-jun

One important use of oxygen is in medicine. People who have trouble breathing are given extra doses of oxygen. In many cases, this "extra oxygen" keeps people alive after they would otherwise have died.

But oxygen has many commercial uses also. The most important use is in the manufacture of metals. More than half of the oxygen produced in the United States is used for this purpose. Oxygen usually ranks third in the list of chemicals produced in the United States each year. In 1996, about 668 billion cubic feet of oxygen was manufactured in the United States. The gas is prepared almost entirely from liquid air.

Discovery and naming

What is air? Ancient peoples thought deeply about that question. And that should not be surprising. It is easy to see how essential air is to many processes. Objects cannot burn without air. Human life cannot survive without air. In fact, ancient peoples thought air must be an "element." But they used the word "element" differently than do modern scientists. To ancient people, an element was something that was very important and basic. Air fit that description, along with fire, water, and earth.

They often thought of air as an element in the modern sense—that it was as simple a material as could be found. Yet, some early scholars believed otherwise. For example, some Chinese scholars, as early as the eighth century A.D., thought of air as having two parts. They called these parts the *yin* and *yang* of air. The properties of the Chinese yin and yang can be compared to the properties of oxygen and **nitrogen.**

The first person in Western Europe to describe the "parts" of air was Italian artist and scientist Leonardo da Vinci (1452–1519). Leonardo pointed out that air is not entirely used up when something is burned in it. He said that air must consist, therefore, of two parts: one part that is consumed in burning and one part that is not.

For many years, Leonardo's ideas were not very popular among scholars. One problem was that early chemists did not have very good equipment. It was difficult for them to collect samples of air and then to study it.

WORDS TO KNOW

Allotropes forms of an element with different physical and chemical properties

Alloy a mixture of two or more metals with properties different from those of the individual metals

Isotopes two or more forms of an element that differ from each other according to their mass number

Periodic table a chart that shows how the chemical elements are related to each other

Radioactive isotope an isotope that breaks apart and gives off some form of radiation

Rusting a process by which a metal combines with oxygen

Joseph Priestley.

In the early 1700s, chemists began to find out more about air, but in a somewhat roundabout way. For example, in 1771 and 1772, Scheele studied the effect of heat on a number of different compounds. In one experiment, he used silver carbonate (Ag_2CO_3), mercury carbonate ($HgCO_3$), and magnesium nitrate ($Mg(NO_3)_2$). When he heated these compounds, he found that a gas was produced. He then studied the properties of that gas. He found that flames burned brightly in the gas. He also found that animals could live when placed in the gas. Without knowing it, Scheele had discovered oxygen. (See sidebar on Scheele in the **chlorine** entry in Volume 1.)

About two years later, Priestley conducted similar experiments by heating mercury oxide (HgO) in a flame. The compound broke down, producing liquid mercury metal and a gas:

$$2HgO \xrightarrow{heat} 2Hg + O_2$$

When Priestley tested the new gas, he found the same properties that Scheele had described.

Antoine-Laurent Lavoisier | French chemist

Antoine-Laurent Lavoisier (1743–94) is often called the father of modern chemistry. He has been given that title for a number of reasons. The most important reason is the explanation he discovered for the process of combustion (burning).

Prior to Lavoisier's research, chemists thought that a burning object gave off a substance to the air. They called that substance phlogiston. When wood burned, for example, chemists said that phlogiston escaped from the wood to the air.

Lavoisier showed that this idea was incorrect. When something burns, it actually combines with oxygen in the air. Combustion, Lavoisier said, is really just oxidation (the process by which something combines with oxygen).

This discovery gave chemists a whole new way to look at chemical changes. The phlogiston theory gradually began to die out. Many of the ideas used in modern chemistry began to develop. No wonder Lavoisier is called the father of this revolution.

Lavoisier led an unusually interesting life. He was an avid chemist who carried out many experiments. But he also had a regular job as a tax collector. His job was to visit homes and businesses and collect taxes. This did not make him a very popular man!

Lavoisier also made some important enemies early in his life. One of these enemies was Jean-Paul Marat (1743–93). Marat thought of himself as a scientist and applied for membership in the French Academy of Scientists. Lavoisier voted against Marat's application. He said that Marat's research was not very good.

Less than a decade later, Lavoisier had reason to regret that decision. Marat had become a leader in the French Revolution (1774-1815). He accused Lavoisier of plotting against the revolution. He also said that Lavoisier was carrying out dangerous secret experiments.

These accusations were not true. But Marat was now a very powerful man. He was able to have Lavoisier convicted of the charges against him. On May 8, 1794, Lavoisier was beheaded and buried in an unmarked grave. Some people have said that Lavoisier's death was the worst single consequence of the French Revolution.

Priestley even tried breathing the new gas he had produced. His description of that experience has now become famous:

> The feeling of it [the new gas, oxygen] to my lungs was not sensibly different from that of common air, but I fancied that my breast felt peculiarly light and

> easy for some time afterwards. Who can tell but that, in time, this pure air may become a fashionable article in luxury? Hitherto only two mice and myself have had the privilege of breathing it.

Some people think Scheele should get credit for discovering oxygen. He completed his experiments earlier than did Priestley. But his publisher was very slow in printing Scheele's reports. They actually came out after Priestley's reports. So most historians agree that Scheele and Priestly should share credit for discovering oxygen.

Neither Scheele nor Priestley fully understood the importance of their discovery. That step was taken by French chemist Antoine-Laurent Lavoisier (1743–94). Lavoisier was the first person to declare that the new gas was an element. He was also the first person to explain how oxygen is involved in burning. In addition, he suggested a name for the gas. That name, oxygen, comes from Greek words that mean "acidic" *(oxy-)* and "forming" *(-gen)*. Lavoisier chose the name because he thought that all acids contain oxygen. Therefore, the new element was responsible for "forming acids." In this one respect, however, Lavoisier was wrong. All acids do not contain oxygen, although some do.

Physical properties

Oxygen is a colorless, odorless, tasteless gas. It changes from a gas to a liquid at a temperature of –182.96°C (–297.33°F). The liquid formed has a slightly bluish color to it. Liquid oxygen can then be solidified or frozen at a temperature of –218.4°C (–361.2°F). The density of oxygen is 1.429 grams per liter. By comparison, the density of air is about 1.29 grams per liter.

Oxygen exists in three allotropic forms. Allotropes are forms of an element with different physical and chemical properties. The three allotropes of oxygen are normal oxygen, or diatomic oxygen, or dioxygen; nascent, atomic, or monatomic oxygen; and ozone, or triatomic oxygen. The three allotropes differ from each other in a number of ways.

First, they differ on the simplest level of atoms and molecules. The oxygen that we are most familiar with in the atmosphere has two atoms in every molecule. Chemists show this by writing the formula as O_2. The small "2" means "two atoms per molecule."

By comparison, nascent oxygen has only one atom per molecule. The formula is simply O, or sometimes (O). The parentheses indicate that nascent oxygen does not exist very long under normal conditions. It has a tendency to form dioxygen:

$$O + O \rightarrow O_2$$

That is, dioxygen is the normal condition of oxygen at room temperature.

The third allotrope of oxygen, ozone, has three atoms in each molecule. The chemical formula is O_3. Like nascent oxygen, ozone does not exist for very long under normal conditions. It tends to break down and form dioxygen:

$$2O_3 \rightarrow 3O_2$$

Ozone does occur in fairly large amounts under special conditions. For example, there is an unusually large amount of ozone in the Earth's upper atmosphere. That ozone layer is important to life on Earth. It shields out harmful radiation that comes from the Sun. Ozone is also sometimes found closer to the Earth's surface. It is produced when gasoline is burned in cars and trucks. It is part of the condition known as air pollution. Ozone at ground level is not helpful to life, and may cause health problems for plants, humans, and other animals.

The physical properties of ozone are somewhat different from those of dioxygen. It has a slightly bluish color as both a gas and a liquid. It changes to a liquid at a temperature of −111.9°C (−169.4°F) and from a liquid to a solid at −193°C (−315°F). The density is 2.144 grams per liter.

Chemical properties

Oxygen's most important chemical property is that it supports combustion. That is, it helps other objects to burn. The combustion (burning) of charcoal is an example. Charcoal is nearly pure carbon (C):

$$C + O_2 \rightarrow CO_2$$

Oxygen also combines with elements at room temperature. Rusting is an example. Rusting is a process by which a metal combines with oxygen. When iron rusts, it combines with oxygen:

$$4Fe + 3O_2 \rightarrow 2Fe_2O_3$$

Oxygen also reacts with many compounds. Decay is an example. Decay is the process by which once-living material combines with oxygen. The products of decay are mainly carbon dioxide (CO_2) and water (H_2O):

$$\text{dead matter} + O_2 \rightarrow CO_2 + H_2O$$

(The chemical formula for "dead matter" is too complicated to use here.)

Oxygen itself does not burn. A lighted match in a container of pure oxygen burns much brighter, but the oxygen does not catch fire.

Occurrence in nature

Oxygen occurs mainly as an element in the atmosphere. It makes up 20.948 percent of the atmosphere. It also occurs in oceans, lakes, rivers, and ice caps in the form of water. Nearly 89 percent of the weight of water is oxygen. Oxygen is also the most abundant element in the Earth's crust. Its abundance is estimated at about 45 percent in the earth. That makes it almost twice as abundant as the next most common element, **silicon.**

Oxygen occurs in all kinds of minerals. Some common examples include the oxides, carbonates, nitrates, sulfates, and phosphates. Oxides are chemical compounds that contain oxygen and one other element. Calcium oxide, or lime or quicklime (CaO), is an example. Carbonates are compounds that contain oxygen, **carbon,** and at least one other element. Sodium carbonate, or soda, soda ash, or sal soda (Na_2CO_3), is an example. It is often found in detergents and cleaning products.

Nitrates, sulfates, and phosphates also contain oxygen and other elements. The other elements in these compounds are nitrogen, sulfur, or **phosphorus** plus one other element. Examples of these compounds are **potassium** nitrate, or saltpeter (KNO_3); magnesium sulfate, or Epsom salts ($MgSO_4$); and calcium phosphate ($Ca_3(PO_4)_2$).

Isotopes

There are three naturally occurring isotope of oxygen: oxygen-16, oxygen-17, and oxygen-18. Isotopes are two or more forms of an element. Isotopes differ from each other according to their mass number. The number written to the right of the element's name is the mass number. The mass number represents the number of protons plus neutrons in the nucleus of an atom of the element. The number of protons determines the element, but the number of neutrons in the atom of any one element can vary. Each variation is an isotope.

Five radioactive isotopes of oxygen are known also. A radioactive isotope is one that breaks apart and gives off some form of radiation. Radioactive isotopes are produced when very small particles are fired at atoms. These particles stick in the atoms and make them radioactive.

None of the radioactive isotopes of oxygen has any commercial use.

Extraction

Oxygen is made from liquid air. Liquid air is made by cooling normal atmospheric air to very low temperatures. As the temperature drops, the gases contained in air turn into liquids. At −182.96°C (−297.33°F), oxygen changes from a gas into a liquid. At −195.79°C (−320.42°F), nitrogen changes from a gas into a liquid. And so on. Eventually, all the gases in air can be made to liquefy (change into a liquid).

But the reverse process also takes place. Suppose liquid air in a container warms up slowly. When its temperature reaches −195.79°C, liquid nitrogen changes back to a gas. A container can be put into place to catch the nitrogen as it boils off the liquid air. When the temperature reaches −182.96°C, oxygen changes from a liquid back to a gas. Another container can be put into place. The escaping oxygen can be collected. Oxygen with a purity of 99.995 percent can be made by this method. It is the only method by which oxygen is made for commercial purposes.

Uses

Many people are familiar with oxygen to help preserve lives. In some cases, people are not able to breathe on their own. Conditions such as emphysema damage the lungs. Oxygen cannot pass through the lungs into the blood stream. One way to treat this condition is to force oxygen into the lungs with a pump.

The same method is used to treat other medical conditions. For example, carbon monoxide poisoning occurs when carbon

People who have trouble breathing use oxygen masks and tanks to help them get the oxygen they need.

monoxide gas gets into the blood stream. Auto exhaust, poorly maintained oil furnaces, and wood fires produce carbon monoxide. The carbon monoxide replaces oxygen in the blood. Cells get carbon monoxide instead of oxygen. But they cannot use carbon monoxide, so they begin to die. Forcing oxygen into the blood can reverse some of the damage. In high enough amounts, it can force the carbon monoxide out of the blood and cells can recover.

Oxygen has other interesting uses. For example, it is used in rocket fuels. It is combined with hydrogen in the rocket

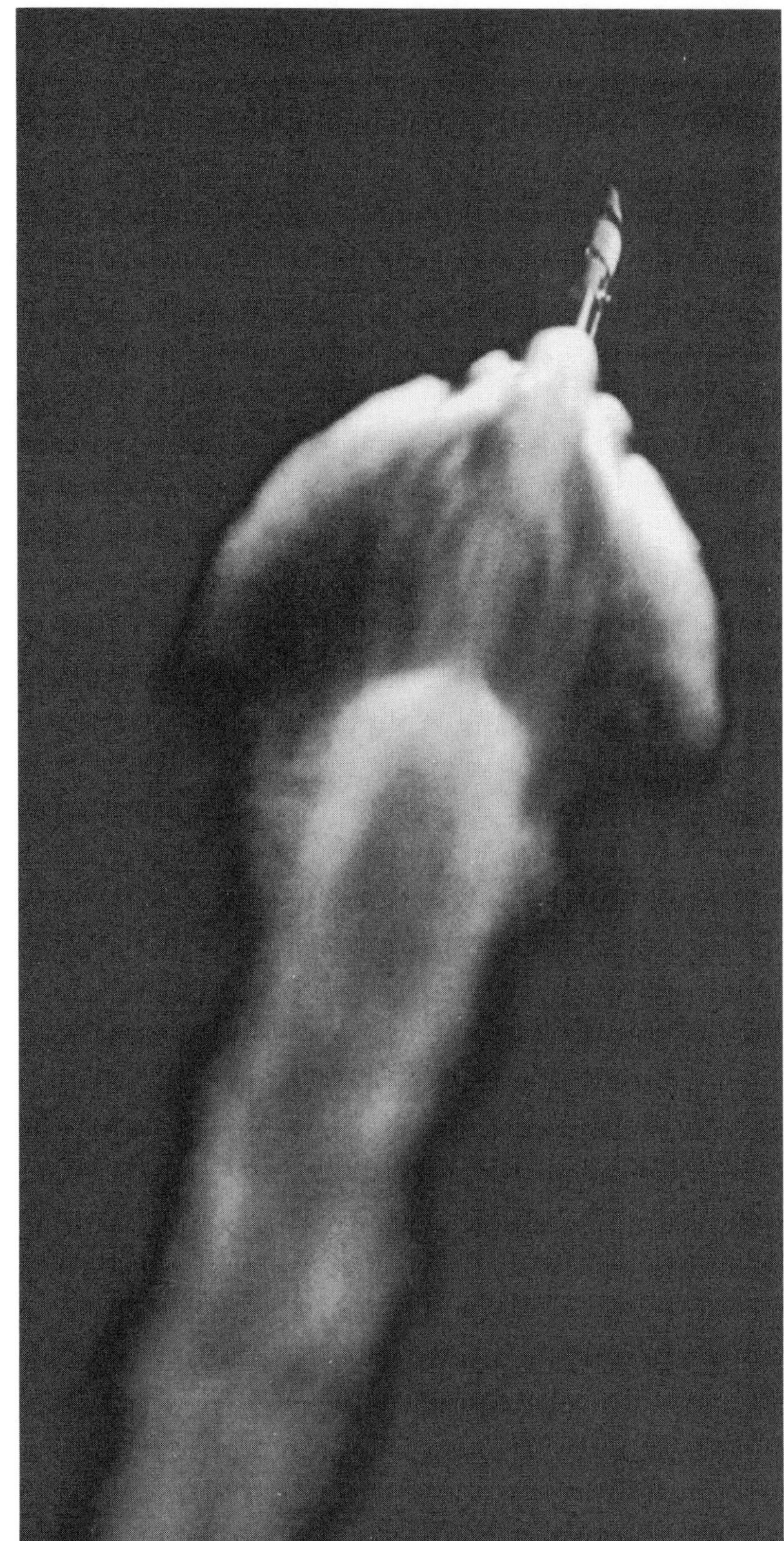

The unmanned Saturn *rocket, shortly after takeoff on January 22, 1968. The combination of oxygen and hydrogen creates enough energy to lift the rocket into space.*

engines. When hydrogen and oxygen combine, they give off very large amounts of energy. The energy is used to lift the rocket into space.

Metal production accounts for the greatest percentage of oxygen use. For example, oxygen is used to burn off carbon and other impurities that are in iron to make steel. A small amount of these impurities may be desirable in steel, but too much makes it brittle and unusable. The carbon and other impurities are burned off in steel-making by blasting oxygen through molten iron.

Two chemical changes that take place during steel-making are shown below:

$$2C + O_2 \rightarrow CO_2$$

$$Si + O_2 \rightarrow SiO_2$$

The carbon dioxide escapes from the steel-making furnace as a gas. The silicon dioxide (SiO_2) forms slag. Slag is a crusty, metallic material that is scraped off after the steel is produced. Other impurities removed by a blast of oxygen are sulfur, phosphorus, **manganese,** and other metals.

Oxygen is also used in the production of such metals as **copper, lead,** and **zinc.** These metals occur in the earth in the form of sulfides, such as copper sulfide (CuS), lead sulfide (PbS), and zinc sulfide (ZnS). The first step in recovering these metals is to convert them to oxides:

$$2CuS + 3O_2 \rightarrow 2CuO + 2SO_2$$

$$2PbS + 3O_2 \rightarrow 2PbO + 2SO_2$$

$$2ZnS + 3O_2 \rightarrow 2ZnO + 2SO_2$$

The oxides are then heated with carbon to make the pure metals:

$$2CuO + C \rightarrow 2Cu + CO_2$$

$$2PbO + C \rightarrow 2Pb + CO_2$$

$$2ZnO + C \rightarrow 2Zn + CO_2$$

Another use of oxygen is in high-temperature torches. The oxyacetylene torch, for example, produces heat by burning acetylene gas (C_2H_2) in pure oxygen. The torch can produce temper-

Coming soon: An oxygen bar near you!

Joseph Priestley was correct! He predicted that "taking a whiff" of oxygen might someday be a luxury for people.

For professional athletes, it already is a common practice. Football players who have run long yardage may inhale some oxygen. It helps them get their energy back.

But now, "taking a whiff" has become even more popular. Oxygen bars are open in Japan, the United States, and other parts of the world. People come to breathe pure oxygen for a few minutes. Of course, they pay a fee for the privilege.

Patrons say that pure oxygen makes them feel better and think more clearly. Others think it improves their looks. Owners of oxygen bars say that city air is often too polluted. They encourage people to "take a whiff" of pure oxygen for their health.

For more information about oxygen bars, contact the National Oxygen Bar Association, 104 North Willow Plaza, Broomfield, CO, 80020; telephone (303) 464-1744; or visit the following website: noba@o2bars.com.

atures of 3,000°C (5,400°F) and cut through steel and other tough alloys.

Oxygen is also used in the chemical industry as a beginning material in making some very important compounds. Sometimes, the steps to get from oxygen to the final compound are lengthy. As an example, ethylene gas (C_2H_4) can be treated with oxygen to form ethylene oxide (CH_2CH_2O):

$$2CH_2H_4 + O_2 \rightarrow 2CH_2H_4O$$

About 60 percent of ethylene oxide produced is made into ethylene glycol ($CH_2CH_2(OH)_2$). Ethylene glycol is used in antifreeze and as a starting point in making polyester fibers, film, plastic containers, bags, and packaging materials.

Compounds

Thousands of oxygen compounds have important commercial uses. Many of these compounds are discussed under other elements.

Health effects

Nearly all organisms require oxygen—bacteria, plants, and animals. Humans, for example, can go weeks and even months without food. They can survive for many days without water.

But they cannot survive more than a few minutes without oxygen.

Oxygen is used by the cells of animal bodies. It is used to "burn" chemicals and produce energy that cells need to stay alive. Without oxygen, cells begin to die in minutes.

BIBLIOGRAPHY

General sources

Print resources

Atkins, P. W. *The Periodic Kingdom: A Journey into the Land of the Chemical Elements.* New York: HarperCollins, 1997.

Budavari, Susan, ed. *The Merck Index.* Rahway, NJ: Merck & Company, Inc., 1989.

Emsley, John, and Jan Neruda. *Elements.* New York: Oxford University Press, 1996.

Greenwood, N. N., and A. Earnshaw. *Chemistry of the Elements.* Oxford: Pergamon Press, 1984.

Heiserman, David L. *Exploring Chemical Elements and Their Compounds.* New York: TAB Books/McGraw Hill, 1991.

Krebs, Robert E. *The History and Use of Our Earth's Chemical Elements: A Reference Guide.* Westport, CT: Greenwood Publishing Group, 1998.

Lewis, Richard J., Sr. *Hawley's Condensed Chemical Dictionary*, 12th edition. New York: Van Nostrand Reinhold Company, 1993.

Nachaef, N. *The Chemical Elements: The Exciting Story of Their Discovery and of the Great Scientists Who Found Them.* Jersey City, NJ: Parkwest Publications, 1997.

Newton, David E. *The Chemical Elements.* New York: Franklin Watts, 1994.

Ruben, Samuel. *Handbook of the Elements.* Chicago: Open Court Publishing Company, 1985.

Stwertka, Albert. *A Guide to the Elements.* New York: Oxford University Press, 1996.

Trifinov, D. N., and V. D. Trifinov. *Chemical Elements: How They Were Discovered.* New York: State Mutual Books, 1985.

Weeks, Mary Elvira, and Henry M. Leicester. *Discovery of the Elements*, 7th edition. Madison, WI: Journal of Chemical Education, 1968.

Internet sources

Readers should be reminded that some Internet sources change frequently. Some of the following web sites may have been removed and new ones added.

Bentor, Yinon. "The Periodic Table of the Elements on the Internet." http://domains.twave.net/domain/yinon/default.html.

Dayah, Michael. "Periodic Table of the Elements." http://www.benray.com/periodic/.

Hyper Chemistry on the Web. "The Periodic Table of Elements." http://tqd.advanced.org/2690/ptable/ptable.html.

KLB Productions. "Yogi's Behemoth Periodic Table of Elements." http://klbproductions.com/yogi/periodic.

Kostas, Tsigaridis. "Periodic Table of the Elements." http://www.edu.uch.gr/~tsigarid/ptoe/info.html.

Los Alamos National Laboratory. Chemical Division. "Periodic Table of the Elements." http://mwanal.lanl.gov/CST/imagemap/periodic/periodic.html.

Phoenix College. "The Pictorial Periodic Table." http://chemlab.pc.maricopa.edu/periodic/periodic.html.

"Taits Periodic Table of the Elements." http://bvsd.co.edu/~stanglt/per/NN3/index.html.

University of Akron. Department of Chemistry. Hardy Research Group. http://ull.chemistry.uakron.edu/periodic_table/.

Winter, Mark. "The Periodic Table on the WWW: Web Elements." http://www.shef.ac.uk/chemistry/web-elements/web-elements-home.html.

Yahoo. "Periodic Table of the Elements." http://www.yahoo.com/science/chemistry/periodic_table_of_the_elements.

Specific elements

Aluminum

Aluminum Association, Inc., 900 19th Street, N.W., Washington, D.C., 20006-7168; (202) 862-5134. Information available on the Internet at www.aluminum.org.

Americium

"Smoke Detectors and Americium." *Nuclear Issues Briefing Paper 35.* Uranium Information Center Ltd., April 1997. Information available on the Internet at http://www.uic.com.au//nip35.htm.

Yu, Jessen, "A:Eeeeeeeee! There are two different types of smoke detectors—photoelectric detectors and ionization detectors." *The Stanford Daily,* May 7, 1996. Information available on the Internet at http://daily.stanford.org/5-7-96/NEWS/index.html.

Bismuth

Bismuth Institute, 301 Borgtstraat – B.1850, Grimbergen, Belgium. Information available on the Internet at http://www.bismuth.be/.

Copper

"Copper: The Red Metal." Information available on the Internet at http://spidergram.ccs.unr.edu/unr/sb204/geology/copper2.html.

The Copper Data Center. Information available on the Internet at http://cdc.copper.org/.

Fluorine

Newton, David E. *The Ozone Dilemma.* Santa Barbara, CA: ABC-CLIO, 1995.

Stille, Darlene R. *Ozone Hole.* Chicago: Childrens Press, 1991.

"The Unofficial Polytetrafluoroethylene (PTFE) Homepage." Information available on the Internet at http://www.net-master.net/~ptfedave/.

Gold

Gold Institute, 1112 16th Street, N.W., Suite 240, Washington, D.C., 20036; (202) 835-0185; e-mail at info@goldinstitute.org. The Institute's web page is available at http://www.goldinstitute.com/about.html.

Hydrogen

American Hydrogen Association, 216 S. Clark, #103, Tempe, AZ 85281; (602) 827-7915.

Canadian Hydrogen Association, 8 King's College, Toronto, Ontario, Canada, M8S 1A4; telephone: (416) 978-2531; fax: (416) 978-0787.

National Hydrogen Association, 1800 M Street, Suite 300, Washington, D.C., 20036-5802; telephone: (202) 223-5547; fax: (202) 223-5537.

Indium

Indium Metal Information Center. Information available on the Internet at http://www.indium.com/metalcenter.html.

Iridium

"Bearing down on the kilogram standard." *Science News,* January 28, 1995, p. 63.

Croft, Sally, "Keeping the kilo from gaining weight." *Science,* May 12, 1995, p. 804.

Ralof, Janet, "Unclogging arteries? Radiation helps." *Science News,* June 14, 1997, p. 364.

Iron

American Iron and Steel Institute, 11017 17th Street, N.W., Washington, D.C., 20036. Information available on the Internet at http://www.steel.org.

Lead

International Lead Zinc Research Organization, 2525 Meridian Parkway, P.O. Box 12036, Research Triangle Park, N.C.; telephone: (919) 361-4647. Information available on the Internet at http://www.ilzro.org/ilzro.html.

Lead Industries Association, 292 Madison Ave., New York, NY 10017.

Lithium

Health Center. "Mood Stabilizer Medications: Lithium." Information available on the Internet at http://www.healthguide.com/Pharmacy/LITHIUM.htm.

Molybdenum

International Molybdenum Association, 7 Hackford Walk, 119-123 Hackford Road, London SW9 0QT. Information available on the Internet at http://www.imoa.org.uk/.

Nickel

Nickel Development Institute, 214 King Street West, Suite 510, Toronto, Ontario, Canada; M5H 3S6; telephone: (416) 591-7999; fax: (416) 591-7987. Information available on the Internet at http://www.nidi.org/.

Nickel Producers Environmental Research Association, 2604 Meridian Parkway, Suite 200, Durham, NC 27713; telephone: (919) 544-7722.

Niobium

"What is Niobium?" Information available on the Internet at http://www.teleport.com/~paulec/Niobium.html.

Yarris, Lynn, "Magnet Sets World Record." *Research Review,* fall 1997.

Palladium

"The Amazing Metal Sponge." Information available on the Internet at http://www.psc.edu/MetaCenter/MetaScience/Articles/Wolf/Wolf.html.

Plutonium

"Electronic Resource Library," Amarillo National Resource Center for Plutonium. Information available on the Internet at http://plutonium-erl.actx.edu/.

Radium

Birch, Beverly. *Marie Curie.* Milwaukee, WI: Gareth Stevens Publishing, 1988.

Keller, Mollie. *Marie Curie.* New York: Franklin Watts, 1982.

Parker, Steve. *Marie Curie and Radium.* New York: HarperCollins, 1992.

Pflaum, Rosalynd. *Grand Obsession: Madame Curie and Her World.* New York: Doubleday, 1989.

Poynter, Margaret. *Marie Curie: Discoverer of Uranium.* Hillside, NJ: Enslow Publishers, 1994.

Saari, Peggy, and Stephen Allison, eds. *Scientists: The Lives and Works of 150 Scientists.* Detroit: U•X•L, 1996, pp. 181–91.

Radon

"A Citizen's Guide to Radon: The Guide to Protecting Yourself and Your Family from Radon." Washington, D.C.: U.S. Environmental Protection Agency, 1992.

Selenium

Selenium-Tellurium Development Association, 11 Broadway, New York, NY 10013. Information available on the Internet at http://www.stda.be/.

Sulfur

Sulfur Institute, 1140 Connecticut Avenue, N.W., Washington, D.C., 20036.

Tellurium

Selenium-Tellurium Development Association, 11 Broadway, New York, NY 10013. Information available on the Internet at http://www.stda.be/.

Tin

Tin Information Center of North America, 1353 Perry Avenue, Columbus, OH 43201.

U.S. Geological Survey, *Minerals Information-1996,* Washington, D.C.: Government Printing Office, 1997.

Uranium

"The Core," The Uranium Institute. Information available on the Internet at http://www.uilondon.org.

Uranium & Nuclear Power Information Centre, Australia. Information available on the Internet at http://www.uic.com.au/index.htm.

Vanadium

Hilliard, Henry E. "Vanadium," *Minerals Yearbook.* Washington, D.C.: U.S. Geological Survey.

McGuire, Rory, "Vanadium battery research recharged by Pinnacle Mining." *Uniken,* June 27, 1997, pp. 6–10. Information available on the Internet at http://www.ceic.unsw.edu.au/centers/vrb/Pinnacle.htm.

Zinc

American Zinc Association, 1112 16th Street, N.W., Washington, D.C., 20036.

International Lead Zinc Research Organization, 2525 Meridian Parkway, P.O. Box 12036, Research Triangle Park, N.C.; telephone: (919) 361-4647. Information available on the Internet at http://www.ilzro.org/ilzro.html.

PICTURE CREDITS

The photographs appearing in *Chemical Elements: From Carbon to Krypton* were reproduced by permission of the following sources:

On the front cover: **Kenneth Eward / BioGrafx, National Audubon Society Collection / Photo Researchers, Inc.:** background image; **Yoav Levy / Phototake NYC:** first box; **Andrew Syred / The National Audubon Society Collection / Photo Researchers, Inc.:** third box.

On the back cover: **Kenneth Eward / BioGrafx, National Audubon Society Collection / Photo Researchers, Inc.:** first box; **Yoav Levy / Phototake NYC:** second box; **Lawrence Migdale / Science Source, National Audubon Society Collection / Photo Researchers, Inc.:** third box.

In the text: **JLM Visuals:** pp. 10, 91, 93, 151, 271, 548, 561; **Library of Congress:** pp. 11, 35, 88, 104, 121, 127, 158, 167, 180, 211, 364, 398, 435, 467, 468, 472, 474, 479, 493; **Field Mark Publications:** pp. 12, 17, 23, 34, 85, 116, 126, 154, 164, 189, 198, 228, 256, 264, 265, 322, 323,

338, 352, 387, 393, 412, 420, 421, 448, 489, 522, 537, 550, 583, 602, 619, 651, 655, 678, 685; **Russ Lappa / Science Source, National Audubon Society Collection / Photo Researchers, Inc.:** pp. 22, 143, 203, 227, 258, 369, 377, 427, 497, 581, 611, 635, 649, 683; **Alexander Tsiaras / National Audubon Society Collection / Photo Researchers, Inc.:** p. 28; **Michael English / Custom Medical Stock Photo, Inc.:** p. 43; **Luis de la Maza / Phototake NYC:** p. 44; **Corbis-Bettmann:** pp. 50, 252, 278, 283, 450, 629; **Mark Elias / AP/Wide World Photos, Inc.:** p. 56; **Earl Scott / National Audubon Society Collection / Photo Researchers, Inc.:** p. 62; **AP/Wide World Photos, Inc.:** pp. 69, 98, 403; **Ray Nelson / Phototake NYC:** p. 70; **Charles D. Winters / National Audubon Society Collection / Photo Researchers, Inc.:** pp. 75, 109, 309, 418, 527, 667; **Department of Energy:** p. 78; **Rich Treptow / National Audubon Society Collection / Photo Researchers, Inc.:** pp. 83, 519; **University of California-Berkeley:** p. 99; **Andrew Mcclenaghan / National Audubon Society Collection / Photo Researchers, Inc.,** p. 105; **Hans Namuth / Photo Researchers, Inc.:** p. 110; **Royal Greenwich Observatory / National Audubon Society Collection / Photo Researchers, Inc.:** p. 124; **Yoav Levy / Phototake NYC:** pp. 129, 140, 320, 353, 665; **Bruce Iverson / Science Photo Library, National Audubon Society Collection/ Photo Researchers, Inc.:** p. 144; **New York Convention & Visitors Bureau, Inc.:** p. 152; **Roland Seitre / BIOS:** p. 156; **U.S. National Aeronautics and Space Administration (NASA):** pp. 160, 404, 441, 607; **Will and Deni McIntrye / National Audubon Society Collection / Photo Researchers, Inc.:** pp. 171, 656; **Erich Schrempp / National Audubon Society Collection / Photo Researchers, Inc.:** p. 172; **Visual Image:** p. 184; **Science Photo Library:** p. 193; **Alfred Pasieka / Science Photo Library, The National Audubon Society Collection / Photo Researchers, Inc.:** pp. 205, 263, 281; **Michael Abbey / Science Source, The National Audubon**

Society Collection / Photo Researchers, Inc.: p. 206; **Dornier Space/Science Photo Library, National Audubon Society Collection / Photo Researchers, Inc.:** p. 213; **Archive Photos, Inc.:** p. 219; **Otto Rogge / Stock Market:** p. 222; **National Center for Atmospheric Research / University Corporation for Atmospheric Research / National Science Foundation 221:** p. 232; **The Granger Collection Ltd., New York:** p. 234; **Federal Aviation Administration:** p. 238; **Ronald Royer / Science Photo Library, National Audubon Society Collection / Photo Researchers, Inc.:** p. 247; **Francois Gohier / Photo Researchers, Inc.:** p. 272; **Frank Rossotto / Stocktrek Photo Agency:** p. 273; **Michael W. Davidson / National Audubon Society Collection / Photo Researchers, Inc.:** p. 295; **Chris Jones / The Stock Market:** p. 302; **Gale Research:** pp. 330, 397; **Spencer Grant / Photo Researchers, Inc.:** p. 331; **Dr. Jeremy Burgess / Science Photo Library, National Audubon Society Collection / Photo Researchers, Inc.:** p. 335; **Kenneth Eward / BioGrafx, National Audubon Society Collection / Photo Researchers, Inc.:** p. 357; **Joshua Roberts / Corbis-Bettmann:** p. 358; **Tom Ives / The Stock Market:** p. 359; **Robert Visser / Greenpeace:** p. 362; **John Tarrell Scott / Galarie Simonne Stern:** p. 371; **Mediafocus:** p. 372; **Franklin Watts:** p. 378; **Philippe Plailly / Science Photo Library, The National Audubon Society Collection / Photo Researchers, Inc.:** pp. 379, 627; **Thomas Del Brase / Stock Market:** p. 388; **Martin Dohrn / National Audubon Society Collection / Photo Researchers, Inc.:** pp. 411; **Eamonn McNulty / National Audubon Society Collection / Photo Researchers, Inc.:** p. 429; **Scott Camazine / National Audubon Society Collection / Photo Researchers, Inc.:** p. 432; **Science Source / National Audubon Society Collection / Photo Researchers, Inc.:** p. 434; **National Archives and Records Administration:** p. 456; **Tony Ward / Photo Researchers, Inc.:** p. 461; **Martin Dohrn / Science Photo Library, National Audubon Soci-**

ety Collection / Photo Researchers, Inc.: p. 475; **Hank Morgan / National Audubon Society Collection / Photo Researchers, Inc.:** p. 482; **David Sutton / Zuma Images / The Stock Market:** p. 504; **Red Elf / Denise Ward-Brown:** p. 507; **Trek Bicycle Corp.:** p. 515; **Adam Hart-Davis / Photo Researchers, Inc.:** p. 530; **Andrew Syred / The National Audubon Society Collection / Photo Researchers, Inc.:** p. 531; **Bryan Peterson / Stock Market:** p. 535; **Lawrence Migdale / Science Source, National Audubon Society Collection / Photo Researchers, Inc.:** pp. 544, 562; **Fireworks by Grucci:** p. 557; **David Taylor / Science Photo Library, National Audubon Society Collection / Photo Researchers, Inc.:** p. 566; **Custom Medical Stock Photo, Inc.:** p. 572; **National Audubon Society Collection / Photo Researchers, Inc.:** 588; **Peter Berndt, M.D. / Custom Medical Stock Photo:** p. 594; **Tony Freeman / PhotoEdit:** p. 600; **Account Phototake / Phototake NYC:** pp. 620, 622; **National Oceanic & Atmospheric Administration:** p. 637; **Paolo Koch / National Audubon Society Collection / Photo Researchers, Inc.:** p. 641; **Department of Energy:** p. 642; **PHOTRI / Stock Market:** p. 643; **Robert L. Wolke:** p. 644; **Rick Altman / The Stock Market:** p. 675; **Archive fur Kunst und Geschichte / Photo Researchers, Inc.:** p. 677.

In the color signatures, volume 1: **Erich Schrempp / National Audubon Society Collection / Photo Researchers, Inc.:** p. 1 (top); **Andrew Mcclenaghan / National Audubon Society collection / Photo Researchers, Inc.:** p. 1 (top); **Charles D. Winters / National Audubon Society Collection / Photo Researchers, Inc.:** pp. 2 (top), 3 (bottom); **Yoav Levy / Phototake NYC:** pp. 2 (bottom), 7 (top), 8; **JLM Visuals:** p. 3 (top); **Lawrence Berkeley Lab / National Audubon Society Collection / Photo Researchers, Inc.:** pp. 4–5; **Earl Scott / National Audubon Society Collection / Photo Researchers, Inc.:** p. 6 (top); **Alexander Tsiaras / National Audubon Society Collection / Photo Researchers, Inc.:** p. 6 (bottom); **Rich Treptow / Nation-**

al Audubon Society Collection / Photo Researchers, Inc.: p. 7 (bottom).

In the color signatures, volume 2: **Chris Jones / The Stock Market:** p. 1 (top); **Philippe Plailly / Science Photo Library, The National Audubon Society Collection / Photo Researchers, Inc.:** p. 1 (bottom); **Chris Rogers / The Stock Market:** p. 2 (top); **Michael W. Davidson / National Audubon Society Collection / Photo Researchers, Inc.:** p. 2 (bottom); **Dornier Space/Science Photo Library, National Audubon Society Collection / Photo Researchers, Inc.:** p. 3 (top); **Robert Visser / Greenpeace:** p. 3 (bottom); **Tom Ives / The Stock Market:** pp. 4–5; **Ronald Royer / Science Photo Library, National Audubon Society Collection / Photo Researchers, Inc.:** p. 6; **Russ Lappa / Science Source, National Audubon Society Collection / Photo Researchers, Inc.:** p. 7 (top); **Thomas Del Brase / Stock Market:** p. 7 (bottom); **Yoav Levy / Phototake NYC:** p. 8.

In the color signatures, volume 3: **JLM Visuals:** p. 1 (top); **Science Source / National Audubon Society Collection / Photo Researchers, Inc.:** p. 1 (bottom); **Scott Camazine / National Audubon Society Collection / Photo Researchers, Inc.:** p. 2; **Department of Energy:** p. 3 (top); **U.S. National Aeronautics and Space Administration (NASA):** p. 3 (bottom); **PHOTRI / Stock Market:** pp. 4–5; **Lawrence Migdale / Science Source, National Audubon Society Collection / Photo Researchers, Inc.:** p. 6 (top); **Bryan Peterson / Stock Market:** p. 6 (bottom); **Charles D. Winters / National Audubon Society Collection / Photo Researchers, Inc.:** p. 7 (top); **Tony Ward / Photo Researchers, Inc.:** p. 7 (bottom); **National Audubon Society Collection / Photo Researchers, Inc.:** p. 8.

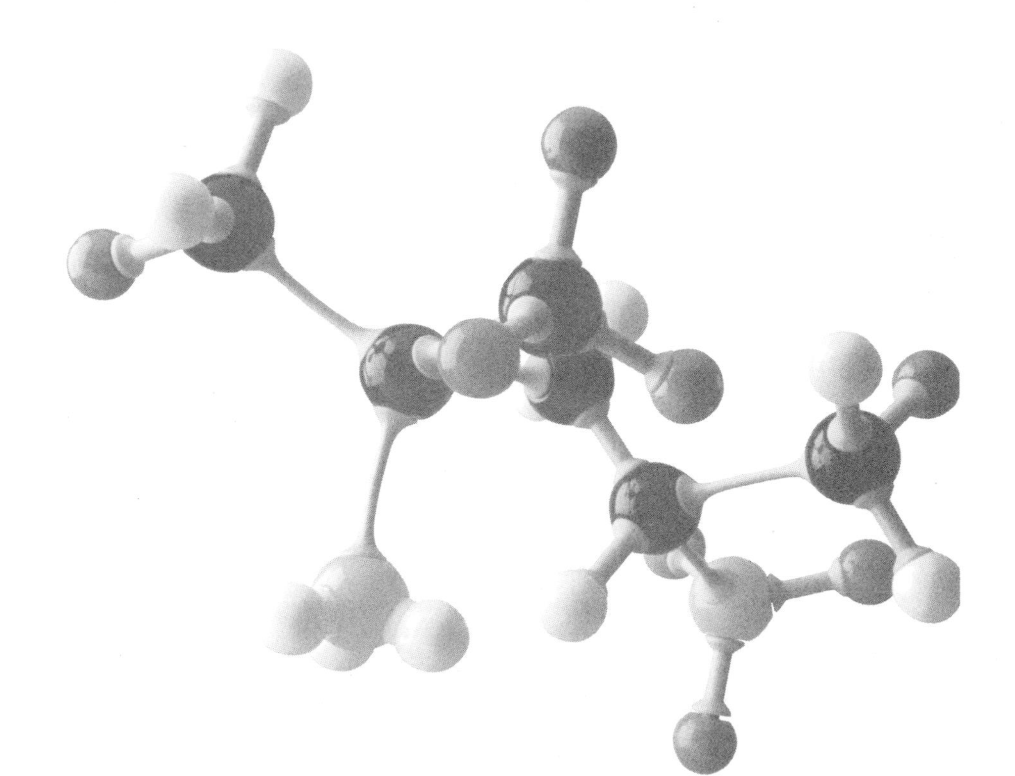

INDEX

A

Abel's metal, *1:* 82
Abelson, Philip H., *2:* 363; *3:* 433
Abrasives, *1:* 71, 117; *3:* 686
Absolute zero, *3:* 666
Ac. *See* Actinium (Ac)
Acetylene gas, *2:* 405
Acids, *2:* 253–54
Actinium (Ac), *1:* **1–3,** 193; *3:* 467
Actinides. *See* Americium (Am); Berkelium (Bk); Californium (Cf); Curium (Cm); Einsteinium (Es); Fermium (Fm); Lawrencium (Lr); Mendelevium (Md); Neptunium (Np); Nobelium (No); Plutonium (Pu); Protactinium (Pa); Thorium (Th); Uranium (U)
Adenosine triphosphate (ATP), *3:* 422
Aerosol cans, *1:* 189 (ill.)
Ag. *See* Silver (Ag)
Agricola, Georgius, *1:* 185
Air, *2:* 382, 396
Airline baggage, *1:* 100
Airport runway lights, *2:* 292
Al. *See* Aluminum (Al)
Alabamine, *1:* 38, 192
Alaska, elements in, *2:* 212, 221, 272, 345; *3:* 674
Albania, elements in, *1:* 137
Albert the Great (Albertus Magnus), *1:* 32
Alchemy, *1:* 19, 20, 31; *3:* 415, 672
Alcohols, *1:* 111
Algae, *3:* 423
Algeria, elements in, *2:* 336
Alka-Seltzer, *3:* 549, 550 (ill.)
Alkali metals. *See* Cesium (Cs); Francium (Fr); Lithium (Li); Potassium (K); Rubidium (Rb); Sodium (Na)
Alkaline earth metals. *See* Barium (Ba); Beryllium (Be); Calcium (Ca); Magnesium (Mg); Radium (Ra); Strontium (Sr)
Alkalis, *3:* 427
Allanite, *2:* 352
Allergies, *2:* 374
Allison, Fred, *1:* 38, 191–92
Alloys. *See* under "Uses" category in most entries
Alpha particles, *2:* 234–35
Alum, *1:* 6
Aluminosis, *1:* 13
Aluminium. *See* Aluminum
Aluminum (Al), *1:* **5–13,** 10 (ill.), 11 (ill.), 12 (ill.), 54; *2:* 310; *3:* 614

Italic type indicates volume number; **boldface** indicates main entries and their page numbers; (ill.) indicates photos and illustrations.

Aluminum family elements. *See* Aluminum (Al); Gallium (Ga); Indium (In); Thallium (Tl)
Aluminum oxide, *1:* 8
Alzheimer's disease, *1:* 13
Am. *See* Americium (Am)
Americium (Am), *1:* **15–18**
Ammonia, *2:* 250, 382, 387–88, 389–90
Ammonium pertechnate, *3:* 577
Andrada e Silva, Jozé Bonifácio de, *2:* 307–8
Andromeda, *3:* 461, 461 (ill.)
Anemia, *2:* 285
Anesthetic, *1:* 89
Anglesite, *2:* 301
Anhydrous ammonia, *2:* 389
Antares (star), *2:* 247 (ill.)
Antimony (Sb), *1:* **19–24,** 22 (ill.)
Antiseptics, *2:* 267
Aqua regia, *1:* 20; *2:* 270, 392; *3:* 426–27, 428, 502, 634–35
Ar. *See* Argon (Ar)
Aragonite, *1:* 41, 90
Arfwedson, Johan August, *2:* 307
Argentina, elements in, *1:* 68
Argentite, *3:* 536
Argon (Ar), *1:* **25–30**
Argon dating, *1:* 28
Argonne National Laboratory, *1:* 38
Argyria, *3:* 539
Argyrodite, *2:* 212
Arizona, elements in, *1:* 151; *3:* 536
Arms, *2:* 341
Arrhenius, Carl Axel, *1:* 161, 169–70; *2:* 241, 313; *3:* 586, 605–6, 660, 664
Arsenic (As), *1:* **31–36,** 142
Arsenopyrite, *1:* 33
Artificial hip shaft, *3:* 572 (ill.)
Artificial pacemakers, *3:* 430, 436
As. *See* Arsenic (As)
Astatine (At), *1:* **37–40**
Asteroids, *2:* 272
At. *See* Astatine (At)
Atom smasher. *See* Particle accelerators
Atomic clocks, *1:* 123–24, 124 (ill.); *3:* 495, 499
Atomic weapons. *See* Nuclear weapons
Atoms, *1:* 103–4
ATP. *See* Adenosine triphosphate
Au. *See* Gold (Au)
Auer, Carl (Baron von Welsbach), *2:* 314, 349, 350; *3:* 453
Audio speakers, *2:* 352 (ill.)
Australia, elements in, *1:* 9, 143, 151, 163; *2:* 204, 221, 280, 301, 309, 329, 370; *3:* 536, 571, 666, 674, 684
Austria, elements in, *2:* 321
Automobile batteries, *3:* 651–52
Avoirdupois system, *2:* 223
Azote, *2:* 383
Azurite, *1:* 151

B

B. *See* Boron (B)
Ba. *See* Barium (Ba)
Babbitt metal, *3:* 614
Bacteria, *1:* 125
Baddeleyite, *2:* 228; *3:* 684
Balard, Antoine-Jérôme, *1:* 73, 74, 76
Ball bearings, *3:* 412 (ill.)
Balloons, *2:* 239, 251, 252–53
Bank vaults, *2:* 331 (ill.)
Barite, *1:* 41, 43, 45, 46–47; *3:* 563
Barium (Ba), *1:* **41–47**
Barrington Crater, *2:* 271 (ill.)
Bartlett, Neil, *2:* 289; *3:* 655
Baryte, *3:* 554
Bastnas, Sweden, *1:* 113, 176; *2:* 294, 350
Bastnasite, *1:* 114, 163, 176; *2:* 196, 294, 351, 352; *3:* 455, 506, 514
Batteries, *1:* 23, 34, 34 (ill.), 81, 85, 85 (ill.), 86; *2:* 303–4, 312, 332, 339, 373; *3:* 462, 651–52
Bauxite, *1:* 9
Bayer process, *1:* 9
Be. *See* Beryllium (Be)
Beach sand, *3:* 599
Becquerel, Antoine-Henri, *3:* 440, 471–72, 478, 479 (ill.)
Belgium, elements in, *1:* 33, 61, 83; *2:* 212; *3:* 520
Beralcast, *1:* 55–56
Berg, Otto, *3:* 485, 486, 576
Berkelium (Bk), *1:* **49–51**
Berthollet; Claude Louis, *2:* 279
Beryl, *1:* 53–54, 54–55, 57
Beryllium (Be), *1:* **53–57**
Berzelius, Jöns Jakob, *1:* 114; *3:* 517–18, 525, 526, 597–98, 618, 682
Beta rays, *3:* 503
Betterton-Kroll process, *1:* 61
Bh. *See* Bohrium (Bh)
Bi. *See* Bismuth (Bi)

Bible, *2:* 218; *3:* 534, 560
Bicycle frames, *3:* 515 (ill.)
Binary compounds, *2:* 229
Biochemistry, *1:* 111
Bipolar disorder, *2:* 311
Bismuth (Bi), *1:* **59–63,**
Bismuth crystals, *1:* 62 (ill.)
Bismuthinite, *1:* 61
Bk. *See* Berkelium (Bk)
Black lung, *1:* 111
Black phosphorus, *3:* 417
Black, Joseph, *2:* 318–19
Bladders, *1:* 36
Bleach, *1:* 126 (ill.), 130
Blimps, *2:* 238 (ill.), 239
Blood, *1:* 29, 138, 155; *2:* 257, 266, 285; *3:* 591
BNCT. *See* Boron neutron capture therapy (BNCT)
Body implants, *3:* 573, 621
Bohr, Niels, *3:* 628
Bohrium (Bh), *3:* 626, 628, 630
Bolivia, elements in, *1:* 21; *3:* 612
"Bologna stone," *1:* 42
Bones, *1:* 72, 96, 100, 160, 187; *2:* 197, 204; *3:* 422, 437, 508, 556–57, 578, 652
Borax, *1:* 65, 67
Boron (B), *1:* **65–72**
Boron neutron capture therapy (BNCT), *1:* 68
Boulangerite, *1:* 21
Bournonite, *1:* 21
Bowels, *2:* 204
Boyle, Robert, *2:* 246
Br. *See* Bromine (Br)
Brackish water, *2:* 263–64
Brain, *1:* 156; *2:* 257, 266, 337; *3:* 419, 578, 657
Brand, Hennig, *3:* 415–17
Brande, William Thomas, *2:* 308
Brandt, Georg, *1:* 141, 142
Brass, *3:* 672, 677
Brauner, Bohuslav, *3:* 460
Brazil, elements in, *1:* 9, 55, 137, 163; *2:* 280, 329, 377; *3:* 571, 612, 661, 665
Breasts, *2:* 204; *3:* 419
Bricks, *3:* 686
Brine, *2:* 264, 321; *3:* 643–44
Bromine (Br), *1:* **73–79,** 75 (ill.)
Bronchi, *2:* 354
Bronchitis, *1:* 57
Bronze, *1:* 150; *3:* 610, 614, 677
Bronze Age, *1:* 150; *3:* 610
Brunetti, Rita, *3:* 460
Buckyballs, *1:* 104, 110
Bunsen, Robert, *1:* 119, 120, 121 (ill.); *2:* 256; *3:* 495, 496, 592

C

C. *See* Carbon (C)
Ca. *See* Calcium (Ca)
Cade, John, *2:* 311
Cadmium (Cd), *1:* **81–86,** 83 (ill.)
Calamine lotion, *1:* 82
Calavarite, *2:* 220
Calcite, *1:* 41, 90
Calcium (Ca), *1:* 27, 61, **88–96;** *3:* 556
Calcium carbonate, *1:* 93 (ill.)
Calcium phosphate, *2:* 401
California, elements in, *1:* 68, 97; *2:* 221, 321, 336; *3:* 555
Californium (Cf), *1:* **97–100,** 99 (ill.)
Calomel, *2:* 333, 340
Canada, elements in, *1:* 83, 121, 143, 151, 163; *2:* 221, 257, 272, 280, 301, 309, 345, 370, 377, 389; *3:* 411, 427, 520, 536, 563, 571, 666, 674
Cancer, *1:* 3, 18, 36, 39, 57, 68, 100, 122, 144, 153, 160; *2:* 204, 222–23, 301, 354; *3:* 420, 437, 476, 480, 482, 503, 508, 556, 587, 667
Capacitors, *3:* 571–72
Carats, *2:* 223
Carbon (C), *1:* 33, **101–112;** *2:* 282–83
Carbon arc lamps, *1:* 89; *2:* 296; *3:* 457, 602
Carbon dioxide, *1:* 105–6, 108; *2:* 405
Carbon dioxide gas, *1:* 105 (ill.)
Carbon family elements. *See* Carbon (C); Germanium (Ge); Lead (Pb); Silicon (Si); Tin (Sn)
Carbon-14 dating, *1:* 107
Carbon monoxide, *1:* 109, 111–12; *2:* 402–3
Carbonates, *2:* 401
Carborundum, *3:* 531
Carnallite, *1:* 128; *2:* 321; *3:* 445, 496
Carnotite, *3:* 641, 649
Carroll, Lewis, *2:* 341
Cassiopeium, *2:* 314
Cassiterite, *3:* 612, 613
Cast iron, *2:* 283
Catalytic convertor, *3:* 430

Catalytic systems, *1:* 116
Cataract surgery, *3:* 504 (ill.)
Cavendish, Henry, *1:* 26; *2:* 245, 246, 383
CCA. *See* Chromated copper arsenate
Cd. *See* Cadmium (Cd)
CD player, *1:* 164 (ill.)
Ce. *See* Cerium (Ce)
Celestine, *3:* 555
Celestite, *3:* 563
Cementation, *1:* 146
Cemented carbide, *3:* 637–38
Centrifuge, *3:* 645
Ceramics, *1:* 23, 115; *2:* 310; *3:* 557, 600
Cerargyrite, *3:* 536
Ceria, *2:* 349; *3:* 453
Cerite, *1:* 114, 176; *2:* 196, 295, 350; *3:* 454, 506, 587
Cerium (Ce), *1:* **113–17;** *2:* 294, 350
Cermets, *3:* 637
Cerussite, *2:* 301
Cesium (Cs), *1:* **119–24**
Cf. *See* Californium (Cf)
CFCs. *See* Chlorofluorocarbons (CFCs)
Chabaneau, Pierre-François, *2:* 270
Chalcocite, *1:* 151
Chalcogens. *See* Oxygen (O); Polonium (Po); Selenium (Se); Sulfur (S); Tellurium (Te)
Chalcopyrite, *1:* 151
Chalk, *1:* 42, 90
Chaptal, Jean Antoine Claude, *2:* 383
Charcoal, *1:* 102, 108, 136; *2:* 400
Chile, elements in, *1:* 33, 151; *2:* 309, 345, 388; *3:* 487, 536
China, elements in, *1:* 21, 33, 43, 55, 61, 68, 83, 186; *2:* 212, 221, 257, 280, 301, 309, 321, 329, 336, 345, 389; *3:* 419, 563, 612, 635, 674
Chinese vermillion, *2:* 336
Chloranthite, *1:* 143
Chlorine (Cl), *1:* **125–33,** 129 (ill.); *2:* 337–38
Chlorofluorocarbons (CFCs), *1:* 132, 183–84, 188–90
Chlorophyll, *2:* 324
Chromated copper arsenate (CCA), *1:* 36
Chromite, *1:* 137
Chromium (Cr), *1:* 57, **135–40,** 140 (ill.); *3:* 648
Chu, Ching-Wu "Paul," *3:* 666
Church organ pipes, *3:* 676 (ill.)
Cigarette lighters, *1:* 116 (ill.)
Cinnabar, *2:* 333, 334, 336, 337; *3:* 563
Circuit board, *2:* 256 (ill.)
Circulatory (blood) system, *1:* 140; *3:* 422
Cl. *See* Chlorine (Cl)
Claude, Georges, *2:* 360
Claus, Carl Ernst (Karl Karlovich Klaus), *3:* 502
Cleve, Per Teodor, *2:* 234, 241, 242; *3:* 605
Cleveite, *2:* 234
Clocks, *1:* 123–24; *3:* 473, 495, 499, 619 (ill.)
Cm. *See* Curium (Cm)
Co. *See* Cobalt (Co)
Coast disease, *1:* 146
Cobalt (Co), *1:* **141–47,** 143 (ill.)
Cobaltite, *1:* 143
Coffinite, *3:* 641
Coins, *3:* 537
Coke, *2:* 282
Colemanite, *1:* 67
Colloidal gold, *2:* 222–23
Colorado, elements in, *1:* 163; *2:* 301, 345
Coloring agents, *1:* 86
Columbian. *See* Niobium
Columbite, *2:* 376–77; *3:* 487
Columbium, *2:* 375, 376
Columbus, Christopher, *2:* 218
Combustion, *2:* 398
Concrete drill bits, *3:* 685 (ill.)
Construction, *1:* 12
Copper (Cu), *1:* **149–56,** 151 (ill.); *2:* 367, 368; *3:* 520, 583, 610
Copying. *See* Photocopying
Coral, *1:* 90, 91 (ill.)
Corrosive sublimate, *2:* 333
Corson, Dale R., *1:* 38
Corundum, *1:* 57
Coster, Dirk, *2:* 226
Courtois, Bernard, *2:* 261–62, 264
Cr. *See* Chromium (Cr)
Crawford, Adair, *3:* 553–54
Cronstedt, Axel Fredrik, *2:* 368
Crookes, Sir William, *3:* 592
Crookesite, *3:* 593
Crustaceans, *1:* 155
Cryolite, *1:* 7, 9, 186
Cs. *See* Cesium (Cs)
Cu. *See* Copper (Cu)
Cuba, elements in, *1:* 143
Cuprite, *1:* 151

Curie, Marie, *1:* 2, 158, 193; *3:* 439, 440, 469, 471, 472 (ill.), 472–73, 476, 478, 597, 598
Curie, Pierre, *1:* 2, 158, 193; *3:* 440, 471, 472–73, 474 (ill.), 478
Curium (Cm), *1:* **97–98,** 157–60
Cyclonium, *3:* 460
Cyclotron, *1:* 38

D

Dating rocks, *1:* 28, 107; *3:* 446–47, 498
Davis-Besse Nuclear Power Plant, *2:* 228 (ill.)
Davy, Sir Humphry, *1:* 6, 41, 66, 87, 88 (ill.), 88–90, 126; *2:* 308, 319, 321; *3:* 443, 444–45, 526, 541, 542, 545, 553, 554, 557
Db. *See* Dubnium (Db)
DDT, *1:* 131
Dead Sea, *1:* 76
Deafness, *2:* 266
Debierne, André, *1:* 2; *3:* 479
Decay, *2:* 401
Del Río, Andrés Manuel, *3:* 646–48
Delaware, elements in, *2:* 321
D'Elhuyard, Don Fausto, *3:* 633, 634
D'Elhuyard, Don Juan José, *3:* 633, 634
Demarçay, Eugène-Anatole, *1:* 175–76; *3:* 506
Dental amalgams, *3:* 539
Dental fillings, *2:* 341, 342
Deodorant, *3:* 686
Dephlogisticated marine acid, *1:* 126
Depression, *2:* 341
Detergents, *3:* 423
Deuterium, *2:* 248
Diabetes, *1:* 140
Diamonds, *1:* 102, 104 (ill)., 105, 107–8; *3:* 686
Dichlorodiphenyltrichloroethane. *See* DDT
Didymia, *2:* 349
Didymium, *2:* 350; *3:* 453, 454, 506
Dilute acids, *3:* 612
Dimethyl mercury, *2:* 340
Dinosaurs, *2:* 269, 272, 272 (ill.)
Dioxygen, *2:* 399, 400
Dirigibles, *2:* 238 (ill.)
Dishware, *3:* 643
Distillation, *2:* 337
DNA, *3:* 678–79
Dolomite, *1:* 42; *2:* 318, 321
Doping, *2:* 211
Doppler radar, *3:* 637 (ill.)
Dorn, Friedrich Ernst, *1:* 2; *3:* 478
Drugs, *1:* 62
Dry ice, *1:* 108, 109 (ill.)
Dry-cell batteries, *2:* 332
Dubna, Russia, *3:* 628
Dubnium (Db), *3:* 626, 628, 630
DuPont Chemical Company, *1:* 188
Dy. *See* Dysprosium (Dy)
Dysprosium (Dy), *1:* **161–64**

E

Ehrlich, Paul, *1:* 35
Einstein, Albert, *1:* 166, 167 (ill.)
Einsteinium (Es), *1:* **165–67**
Ekaboron, *3:* 512
Ekeberg, Anders Gustaf, *3:* 569–70
Electric cars, *1:* 12; *3:* 651–52
Electrical fuses, *3:* 678 (ill.)
Electricity, *1:* 8
Electrolysis, *3:* 443, 537
Electrolytic refining, *2:* 302
Electroplating, *1:* 84–85, 139; *2:* 373; *3:* 614
Emanation, *3:* 478
Emphysema, *2:* 402
Enemas, *1:* 46
Engines, *1:* 116
"Enrich" bulb, *2:* 354
Enzymes, *1:* 146, 156; *2:* 324, 332; *3:* 567
Epsom salts, *2:* 318, 401
Epsomite, *2:* 321
Er. *See* Erbium (Er)
Erbia, *1:* 170; *2:* 242; *3:* 513, 586, 605
Erbium (Er), *1:* **169–73;** *3:* 660, 664
Erythronium, *3:* 648
Es. *See* Einsteinium (Es)
Etching, *1:* 185
Ethylene oxide, *2:* 406
Eu. *See* Europium (Eu)
Europium (Eu), *1:* **175–78**
Euxenite, *3:* 587
Explosives, *2:* 388
Eye glasses, *3:* 622 (ill.)
Eyes, *1:* 24, 29; *2:* 244, 354, 394, 419; *3:* 503, 557

F

F. *See* Fluorine (F)
Fajans, Kasimir, *3:* 466–67
Fe. *See* Iron (Fe)
Federal Helium Program, *2:* 237

Fermi, Enrico, *1:* 179, 180, 180 (ill.)
Fermium (Fm), *1:* **179–81;** *3:* 625
Ferrochromium, *1:* 138
Ferromanganese, *2:* 331
Ferrosilicon, *2:* 322; *3:* 529–30
Ferrotungsten, *3:* 636
Ferrovanadium, *3:* 650
Fertilizers, *2:* 382, 388 (ill.), 388–90; *3:* 420 (ill.), 424–25, 447, 449, 565
"Fiesta Ware," *3:* 643
Film, *3:* 537–38
Finland, elements in, *1:* 137; *2:* 336
Fire extinguishers, *3:* 450 (ill.)
Fire sprinkler systems, *1:* 84
Fireworks, *2:* 323; *3:* 557, 557 (ill.)
Fishing tackle, *1:* 23 (ill.)
Flame retardant materials, *1:* 24, 73, 77
Flash bulbs, *2:* 322–23
Florentium, *3:* 460
Florida, elements in, *2:* 321; *3:* 419, 599, 661
Fluorescent lamps, *2:* 338, 338 (ill.)
Fluoridation, *1:* 187–88
Fluorides, *1:* 183, 184 (ill.)
Fluorine (F), *1:* **183–90**
Fluorspar, *1:* 185; *3:* 506
Fly ash, *3:* 649–50
Fm. *See* Fermium (Fm)
Food irradiation, *1:* 123, 145
"Fool's gold," *3:* 559
Fort Knox, Kentucky, *2:* 224
Fossil fuels, *2:* 253
Foundry molds, *3:* 686
Fourcroy, Antoine François de, *1:* 136
Fr. *See* Francium (Fr)
France, elements in, *2:* 204
Francium (Fr), *1:* **191–94**
Franklinite, *3:* 673
Frasch, Herman, *3:* 564
Fraunhofer, Joseph von, *2:* 256; *3:* 592
Frozen foods, *2:* 386, 387 (ill.)
Fuel cells, *3:* 588–89
Fuller, Buckminster, *1:* 110, 110 (ill.)
Fullerene, *1:* 110
Fungicides, *1:* 131
Fusion, *2:* 247–48
Fusion bombs, *2:* 249

G

Ga. *See* Gallium (Ga)
Gabon, elements in, *2:* 329
Gadolin, Johan, *1:* 161–62; *2:* 195, 242; 313–14; *3:* 586, 606, 663, 664
Gadolinia, *2:* 196
Gadolinite, *1:* 170, 176; *2:* 196, 241; *3:* 487, 514, 587
Gadolinium (Gd), *2:* **195–99**
Gahn, Johann Gottlieb, *2:* 327, 328; *3:* 517–18
Galena, *2:* 301; *3:* 563
Gallium (Ga), *1:* 34, *2:* **201–7,** 203 (ill.)
Gallium arsenide, *2:* 205–7
Galvanizing, *3:* 671, 677
Gamma radiation, *3:* 608
Gamma rays, *3:* 661
Garnet, *3:* 664
Garnierite, *2:* 369
Gas lanterns, *3:* 600, 602 (ill.)
Gasoline, *1:* 77–79; *2:* 303
Gaspra (asteriod), *2:* 273 (ill.)
Gastrointestinal (GI) system, *1:* 42
Gay-Lussac, Joseph Louis, *1:* 66, 69
Gd. *See* Gadolinium (Gd)
Ge. *See* Germanium (Ge)
Geiger counter, *3:* 482 (ill.)
Gemstones, *1:* 57
General Electric Company, *2:* 354
Geodesic dome, *1:* 110
Geoffrey, Claude-Françoise, *1:* 60
Germanite, *2:* 212
Germanium (Ge), *1:* 34; *2:* **209–15**
Germany, elements in, *1:* 83; *2:* 204, 280, 388; *3:* 487, 520, 571
Ghiorso, Albert, *1:* 15, 97, 98 (ill.), 157, 165, 179
Giant Tasmanian lobster, *1:* 156 (ill.)
Giesel, Friedrich O., *1:* 2
Glass, *1:* 23, 71, 115; *2:* 310, 328, 354; *3:* 521–22, 531, 542, 557
Glass bulbs, *1:* 92–93
Glazes, *1:* 20
Glucinium, *1:* 54
Göhring, O. H., *3:* 466–67
Goiter, *2:* 266, 267
Gold (Au), *1:* 20; *2:* **217–24,** 222 (ill.); *3:* 409–10, 415, 537, 580, 581
Gold Rush, *2:* 219
Goodyear, Charles, *3:* 565
Graphite, *1:* 102, 107–8; *2:* 343–44; *3:* 417
Grease, *2:* 311
Greece, elements in, *2:* 218
Greenland spar, *1:* 186
Greenockite, *1:* 83
Gregor, William, *3:* 617, 618

Group 1 (IA) family elements. *See* Cesium (Cs); Francium (Fr); Hydrogen (H); Lithium (Li); Potassium (K); Rubidium (Rb); Sodium (Na)
Group 2 (IIA) family elements. *See* Barium (Ba); Beryllium (Be); Calcium (Ca); Magnesium (Mg); Radium (Ra); Strontium (Sr)
Group 3 (IIIB) family elements. *See* Actinium (Ac); Lanthanum (La); Scandium (Sc); Yttrium (Y)
Group 4 (IVB) family elements. *See* Hafnium (Hf); Rutherfordium (Rf); Titanium (Ti); Zirconium (Zr)
Group 5 (VB) family elements. *See* Dubnium (Db); Niobium (Nb); Tantalum (Ta); Vanadium (V)
Group 6 (VIB) family elements. *See* Chromium (Cr); Molybdenum (Mo); Seaborgium (Sg); Tungsten (W)
Group 7 (VIIB) family elements. *See* Bohrium (Bh); Manganese (Mn); Rhenium (Re); Technetium (Tc)
Group 8 (VIIIB) family elements. *See* Hassium (Hs); Iron (Fe); Osmium (Os); Ruthenium (Ru)
Group 9 (VIIIB) family elements. *See* Cobalt (Co); Iridium (Ir); Meitnerium (Mt); Rhodium (Rh)
Group 10 (VIIIB) family elements. *See* Nickel (Ni); Palladium (Pd); Platinum (Pt); Ununnilium (Uun)
Group 11 (IB) family elements. *See* Copper (Cu); Gold (Au); Silver (Ag); Unununium (Uuu)
Group 12 (IIB) family elements. *See* Cadmium (Cd); Mercury (Hg); Ununbiium (Uub); Zinc (Zn)
Group 13 (IIIA) family elements. *See* Aluminum (Al); Boron (B); Gallium (Ga); Indium (In); Thallium (Tl)
Group 14 (IVA) family elements. *See* Carbon (C); Germanium (Ge); Lead (Pb); Silicon (Si); Tin (Sn)
Group 15 (VA) family elements. *See* Antimony (Sb); Arsenic (As); Bismuth (Bi); Phosphorus (P)
Group 16 (VIA) family elements. *See* Oxygen (O); Polonium (Po); Selenium (Se); Sulfur (S); Tellurium (Te)
Group 17 (VIIA) family elements. *See* Astatine (At); Bromine (Br); Chlorine (Cl); Fluorine (F); Iodine (I); Nitrogen (N)
Group 18 (VIIIA) family elements. *See* Argon (Ar); Helium (He); Krypton (Kr); Neon (Ne); Radon (Rn); Xenon (Xe)
Guinea, elements in, *1:* 9
Gums, *2:* 341
Gypsum, *1:* 92

H

H. *See* Hydrogen (H)
Haber, Fritz, *2:* 387–88
Hafnium (Hf), *2:* **225–29,** 227 (ill.)
Hahn, Otto, *3:* 467, 468 (ill.), 469
Hahnium. *See* Dubnium
Half lives, *2:* 365. *See also* under "Isotopes" category in each entry
Halite, *1:* 128; *3:* 543
Hall, Charles Martin, *1:* 7
Halogens. *See* Astatine (At); Bromine (Br); Chlorine (Cl); Fluorine (F); Iodine (I)
Hassium (Hs), *3:* 626, 629, 630
Hatchett, Charles, *2:* 375–76; *3:* 569, 570
Hatmakers, *2:* 341
Haüy, René-Just, *1:* 54, 136
Hayyan, Abu Musa Jabir Ibn, *1:* 20
HDTV. *See* High-definition television (HDTV)
He. *See* Helium (He)
Heart scan, *3:* 594 (ill.)
Heart, *1:* 24; *2:* 257; *3:* 591, 593
Heat exchange medium, *3:* 546
Helicopters, *1:* 56 (ill.)
Helium (He), *1:* 49–50, 97–98; *2:* **231–39**
Hematite, *2:* 280
Hemocyanin, *1:* 155
Hemoglobin, *1:* 155; *2:* 285
Hevesy, George Charles de, *2:* 226
Hf. *See* Hafnium (Hf)
Hg. *See* Mercury (Hg)
High-definition television (HDTV), *1:* 173
Hindenburg, 2: 251, 252 (ill.), 253
Hippocrates, *1:* 34
Hisinger, Wilhelm, *1:* 114
Historic documents, *2:* 386
Hitler, Adolf, *1:* 180; *2:* 251; *3:* 469
Hjelm, Peter Jacob, *2:* 343, 344
Ho. *See* Holmium (Ho)
Hohenheim, Theophrastus Bombastus von, *3:* 672
Holmia, *2:* 242

Holmium (Ho), *2:* **241–44**
Horseshoe magnet, *2:* 281 (ill.)
Hs. *See* Hassium (Hs)
Hungary, elements in, *3:* 580
Hutchinsonite, *3:* 593
Hydrated copper carbonate, *1:* 150–51, 152
Hydrocarbons, *1:* 110
Hydrochloric acid, *1:* 126, 132
Hydrofluoric acid, *1:* 185
Hydrogen (H), *2:* **245–54**
Hydrogen bomb, *1:* 165, 179; *2:* 249
Hydrogen chloride, *1:* 128
Hydrogen economy, *2:* 253
Hydrogen sulfide, *3:* 562
Hydrogenation, *2:* 251–52
Hypertension, *3:* 551

I

I. *See* Iodine (I)
Idaho, elements in, *1:* 21, 163; *2:* 301, 345; *3:* 419, 536
Illinium, *3:* 460
Illinois, elements in, *2:* 301
Ilmenite, *3:* 618, 619
In. *See* Indium (In)
Incandescent lights, *3:* 637
India, elements in, *1:* 43, 137, 163; *2:* 218, 280, 389; *3:* 661
Indium (In), *2:* **255–59,** 258 (ill.)
Indonesia, elements in, *1:* 151; *2:* 370, 389; *3:* 612
Inert gases. *See* Argon (Ar); Helium (He); Krypton (Kr); Neon (Ne); Radon (Rn); Xenon (Xe)
Inner ear, *2:* 198
Insecticides, *1:* 131; *3:* 565
Institute for Heavy Ion Research, *3:* 626–27, 629
Intelsat VI (satellite), *3:* 607 (ill.)
International Bureau of Weights and Measures, *2:* 274
International Union of Pure and Applied Chemistry (IUPAC), *3:* 628, 629
Intestines, *1:* 140; *3:* 422, 551
Iodine (I), *1:* 39; *2:* **261–67**
Iodine crystal, *2:* 263 (ill.)
Ir. *See* Iridium (Ir)
Iran, elements in, *3:* 555
Iraq, elements in, *1:* 150
Iridium (Ir), *2:* **269–75,** 392
Iridosmine, *2:* 272; *3:* 492
Iron (Fe), *1:* 22; *2:* **277–85,** 327, 328; *3:* 613
Iron Age, *1:* 150; *2:* 277
Iron stoves, *2:* 283 (ill.)
Irradiation. *See* Food irradiation
Israel, elements in, *1:* 73; *3:* 419
Itai-itai disease, *1:* 86
Italy, elements in, *2:* 218

J

Jacinth, *3:* 682
Jamaica, elements in, *1:* 9
James, Charles, *1:* 170; *2:* 314; *3:* 606
James, Ralph A., *1:* 15, 157
Jamesonite, *1:* 21
Janssen, Pierre, *2:* 231, 233
Japan, elements in, *1:* 61, 83; *2:* 280; *3:* 520, 563
Jargon of Ceylon, *3:* 682
Jewelry, *2:* 224, 380; *3:* 428, 537, 685
Joint Institute of Nuclear Research, *3:* 626–27, 628
Joints, *1:* 163
Jordan, elements in, *3:* 419

K

K. *See* Potassium (K)
Kaim, Ignatius Gottfried, *2:* 328
Kamerlingh-Onnes, Heike, *3:* 666
Kansas, elements in, *2:* 236
Kazakhstan, elements in, *1:* 43, 83, 137; *2:* 329
Kelp, *2:* 264, 264 (ill.)
Kentucky, elements in, *1:* 9; *2:* 224
Kernite, *1:* 67
Kesterson Reservoir, *3:* 523
Kidneys, *1:* 24, 36, 140, 156; *2:* 204, 266, 337, 341, 342; *3:* 578, 652
Kilogram, *2:* 274
Kirchhoff, Gustav Robert, *1:* 119, 120; *2:* 256; *3:* 495, 496, 592
Klaproth, Martin Heinrich, *1:* 114; *3:* 580, 618, 640, 681, 682
Klaus, Karl Karlovich, 502
Knife, *3:* 651 (ill.)
"Knocking," *1:* 77; *2:* 303
Kr. *See* Krypton (Kr)
Krypton (Kr), *2:* **287–92**
Kryptonite, *2:* 289
Kyrgyzstan, elements in, *1:* 21; *2:* 336

L

La. *See* Lanthanum (La)
Laboratory vessel, *1:* 75 (ill.)

Lamps, *1:* 89; *2:* 292, 296, 338–39; *3:* 516, 657
Lamy, Claude-Auguste, *3:* 592
Langbeinite, *3:* 445
Langlet, Nils Abraham, *2:* 234
Lanthanum (La), *2:* **293–97,** 350; *3:* 454
Lanthanum aluminate crystal, *2:* 295 (ill.)
Lanthanides. *See* Cerium (Ce); Dysprosium (Dy); Erbium (Er); Europium (Eu); Gadolinium (Gd); Holmium (Ho); Lutetium (Lu); Neodymium (Nd); Praseodymium (Pr); Promethium (Pm); Samarium (Sm); Terbium (Tb); Thulium (Tm); Ytterbium (Yb)
Laptop computers, *2:* 372 (ill.)
Las Vegas, Nevada, *2:* 359 (ill.), 360
Laser printers, *3:* 583 (ill.)
Lasers, *1:* 28 (ill.), 29, 115, 171 (ill.), 173; *2:* 206, 243–44, 353; *3:* 508, 608, 661–62, 664, 668
Laughing gas, *1:* 89
Lavoisier, Antoine-Laurent, *1:* 88; *2:* 246, 383, 398 (ill.), 398–99
Lawrence Berkeley Laboratory (LBL) at the University of California at Berkeley, *2:* 351, 380; *3:* 626–27
Lawrence, Ernest O., *3:* 628, 629 (ill.)
Lawrencium (Lr), *3:* 626, 628, 630
Le Blanc, Nicolas, *3:* 447
Lead (Pb), *1:* 23, 34, 62; *2:* **299–305,** 344; *3:* 583
Lead-antimony alloys, *1:* 23 (ill.)
Lead canisters, *3:* 656 (ill.)
Lead poisoning, *2:* 305
Lead smelting, *2:* 302 (ill.)
Lead storage battery, *1:* 34 (ill.); *2:* 303–4
Leaded gasoline, *1:* 77–79; *2:* 303
Leak detection, *2:* 239, 290; *3:* 481–82, 545
Lecoq de Boisbaudran, Paul Émile, *1:* 162; *2:* 202; *3:* 506
Lee, David M., *2:* 232
Legs, *2:* 341
Lehmann, Johann Gottlob, *1:* 136; *3:* 634
Lemery, Nicolas, *1:* 20
Leonardo da Vinci, *2:* 396
Lepidolite, *1:* 121; *2:* 309; *3:* 496
Li. *See* Lithium (Li)
Lichtenberg's metal, *1:* 82
Light bulbs, *1:* 28–29; *2:* 353 (ill.), 354, 386; *3:* 546–47
Light-emitting diodes (LEDs), *1:* 34; *2:* 205 (ill.), 205–6
Lighter-than-air balloons, *2:* 251, 252–53
Lime, *1:* 94–96
Limestone, *1:* 42, 87–88, 90, 92, 93; *2:* 282
Limonite, *2:* 280
Linnaeite, *1:* 143
Lipinski, Tara, *2:* 219 (ill.)
Lipowitz's metal, *1:* 82
Liquid air, *2:* 236, 287, 288, 385, 402; *3:* 654
Liquid nitrogen, *2:* 386–87
Liquid oils, *2:* 252
Lithium (Li), *2:* **307–12,** 309 (ill.)
Lithium carbonate, *2:* 310–11
Liver, *1:* 24, 36, 156; *2:* 204, 222, 257, 266; *3:* 422, 578
Lorandite, *3:* 593
Louisiana, elements in, *1:* 44
Louyet, Paulin, *1:* 185
Löwig, Carl, *1:* 73, 74, 76
Lr. *See* Lawrencium (Lr)
Lu. *See* Lutetium (Lu)
Lubricant, *2:* 380
Luminescence, *3:* 462–63
Lungs, *1:* 24, 36; *2:* 266, 285; *3:* 480, 652, 657
Lutecium, *2:* 314
Lutetium (Lu), *2:* **313–16**

M

Mackenzie, Kenneth R., *1:* 38
McMillan, Edwin M., *2:* 362–63, 364 (ill.); *3:* 433
McVeigh, Timothy, *2:* 389
Mad Hatter, *2:* 341
Magic numbers, *3:* 631
Magnesia, *2:* 328
Magnesite, *1:* 42; *2:* 321
Magnesium (Mg), *1:* 27, 61; *2:* **317–25,** 320 (ill.)
Magnetic field, *1:* 144 (ill.), 145
Magnetite, *2:* 280
Magnets, *1:* 69 (ill.), 70, 70 (ill.); *2:* 354, 379 (ill.), 380; *3:* 508, 665 (ill.)
Malachite, *1:* 151, 155
Manganese (Mn), *2:* **327–32**
Manganin, *2:* 331
Manganite, *2:* 329

Manic-depressive disorder, *2:* 311
Marat, Jean-Paul, *2:* 398
Marble, *1:* 42, 90
Marggraf, Andreas Sigismund, *1:* 6
Marignac, Jean-Charles-Galissard de, *2:* 196, 314; *3:* 506, 586, 659–60
Mariner 10 (space probe), *3:* 441 (ill.)
Mars, *3:* 655
Mars Pathfinder, 1: 159, 160 (ill.)
Masurium, *3:* 486, 576
Matches, *3:* 421 (ill.), 422
Md. *See* Mendelevium (Md)
Meitner, Lise, *3:* 467, 467 (ill.), 469, 629
Meitnerium (Mt), *3:* 626, 629, 630
Mendeleev, Dmitri, xxiv; *1:* 37; *2:* 201–2, 209–10; *3:* 459, 511, 512, 513, 628
Mendelevium (Md), *3:* 626, 628, 630
Mental retardation, *2:* 266
Mercury (Hg), *1:* 6, 32; *2:* 223, **333–42,** 335 (ill.)
Metal fatigue, *3:* 583
Metallurgical Research Laboratory (MRL), *1:* 157
Meteorites, *2:* 271–72, 278, 369; *3:* 528
Meteors, *2:* 269
Meter, *2:* 291
Methane gas, *2:* 382
Methanol, *2:* 250–51
Methyl alcohol, *1:* 110–11
Mexico, elements in, *1:* 33, 61, 186; *2:* 301; *3:* 536, 555, 563, 674
Meyer, Lothar, xxiv
Mg. *See* Magnesium (Mg)
Michigan, elements in, *2:* 321; *3:* 447
Microwave ovens, *2:* 198 (ill.), 199
Midgley, Thomas, Jr., *1:* 188
Milk, *1:* 96
Mimetite, *2:* 301
"Mineral alkali," *3:* 444
Miner's lamp, *1:* 89
Misch metal, *1:* 116, 116 (ill.); *2:* 296; *3:* 456–57
Missouri, elements in, *2:* 301; *3:* 674
Mn. *See* Manganese (Mn)
Mo. *See* Molybdenum (Mo)
Mohs scale, *1:* 103; *2:* 308; *3:* 528, 673
Moissan, Henri, *1:* 66, 183, 185, 187
Molybdenite, *2:* 343; *3:* 487
Molybdenum (Mo), *2:* **343–47;** *3:* 578
Monazite, *1:* 114, 162, 163, 176; *2:* 196, 241, 294, 315, 351, 352; *3:* 455, 506, 514, 587, 599, 600, 661, 665–66
Monge, Gaspard, *2:* 279
Mongolia, elements in, *1:* 186
Montana, elements in, *1:* 9, 151, 163; *2:* 221, 301; *3:* 428, 514, 674
Moon, *2:* 233; *3:* 665
Morgan, Leon O., *1:* 15
Morocco, elements in, *1:* 43; *3:* 419
Mosander, Carl Gustav, *1:* 113, 170; *2:* 293, 294, 350; *3:* 453, 454, 586, 664
Mouth, *2:* 341
Mt. *See* Meitnerium (Mt)
Müller von Reichenstein, Baron Franz Joseph, *3:* 579, 580
Murder, *1:* 35–36
Muscles, *2:* 341; *3:* 551
Mussolini, Benito, *1:* 180
Mutism, *2:* 266
Muwaffaw, Abu Mansur, *1:* 89

N

Na. *See* Sodium (Na)
Namibia, elements in, *1:* 33
Nanotubes, *1:* 110
Napoleon Bonaparte, *1:* 66
Nascent oxygen, *2:* 400
Natrium, 3: 542
Natron, 3: 542
Natural gas, *2:* 236
Nb. *See* Niobium (Nb)
Nd. *See* Neodymium (Nd)
Ne. *See* Neon (Ne)
Nebraska, elements in, *1:* 61
Nematicides, *1:* 131
Neodymium (Nd), *2:* **349–54;** *3:* 454
Neodymium-iron-boron (NIB) magnet, *2:* 354
Neodymium-yttrium aluminum garnet (Nd:YAG) laser, *2:* 353–54; *3:* 668–69
Neon (Ne), *2:* **355–60**
Neon atom, *2:* 357 (ill.)
Neon lights, *2:* 292, 358 (ill.), 359–60; *3:* 657
Neoytterbium, *2:* 314
Neptune, *2:* 363
Neptunium (Np), *2:* **361–66**
Nervousness, *2:* 341
Netherlands, elements in, *2:* 389
Neutron radiography, *2:* 197
Nevada, elements in, *1:* 44, 151; *2:* 221, 309, 321, 336, 345; *3:* 536

New Caledonia, elements in, *2:* 370
New Mexico, elements in, *1:* 151; *2:* 345; *3:* 447
New York, elements in, *2:* 301; *3:* 674
Newton's metal, *1:* 82
Ni. *See* Nickel (Ni)
Nicad batteries. *See* Nickel-cadmium (nicad) batteries
Nichols, Terry, *2:* 389
Nickel (Ni), *2:* **367–74,** 369 (ill.)
Nickel allergy, *2:* 374
Nickel-cadmium (nicad) batteries, *1:* 81, 85, 85 (ill.); *2:* 372 (ill.), 373
Nielsbohrium. *See* Bohrium
Nilson, Lars Fredrik, *3:* 511, 512–13, 659, 660
Niobium (Nb), *2:* **375–80, 377 (ill.);** *3:* 569–70
Nitrates, *2:* 384, 388, 401
Nitric acid, *2:* 390; *3:* 429–30
Nitrogen (N), *1:* 27; *2:* **381–90**
Nitrogen family elements. *See* Antimony (Sb); Bismuth (Bi); Nitrogen (N); Phosphorus (P)
Nitrogen fixation, *2:* 387
Nitrous oxide gas, *1:* 89
No. *See* Nobelium (No)
Nobel, Alfred, *3:* 628
Nobelium (No), *3:* 626, 628, 630
Noble gases. *See* Argon (Ar); Helium (He); Krypton (Kr); Neon (Ne); Radon (Rn); Xenon (Xe)
Noddack, Walter, *3:* 485, 486, 576
Nodules, *2:* 329
North Carolina, elements in, *1:* 9, 163; *2:* 309, 321; *3:* 419
North Korea, elements in, *2:* 321
Np. *See* Neptunium (Np)
Nuclear chain reaction, *3:* 644 (ill.)
Nuclear explosion, *3:* 643 (ill.)
Nuclear fission, *1:* 67, 122, 176; *2:* 198, 226–27, 361; *3:* 431, 436, 469, 546, 599, 639, 644–45, 682
Nuclear power plants, *1:* 108; *2:* 198, 225, 361, 379–80; *3:* 436, 448, 546, 598, 639, 645, 682
Nuclear weapons, *3:* 436, 556, 598, 682
Nucleic acids, *3:* 422

O

O. *See* Oxygen (O)
Oak Ridge Laboratory, *3:* 460
Ocean Song (sculpture), *2:* 371 (ill.)
Oersted, Hans Christian, *1:* 6
Ohio, elements in, *1:* 55
Oil drilling, *1:* 45–46, 78 (ill.)
Oil industry, *3:* 429
Oil pipelines, *1:* 21–22, 122; *3:* 545
Oklahoma, elements in, *2:* 236; *3:* 514
Oklahoma City bombing, *2:* 389
"Old Nick's copper," *2:* 367
Olympic gold medal, *2:* 218, 219 (ill.)
Open-heart surgery, *3:* 411 (ill.)
Optical fibers, *1:* 169, 172 (ill.), 173; *2:* 214, 296–97
Oregon, elements in, *1:* 9; *2:* 370
Organ pipes, *3:* 611, 676 (ill.)
Organic chemistry, *1:* 101
Orpiment, *1:* 33
Orthite, *3:* 506
Os. *See* Osmium (Os)
Osann, Gottfried W., *3:* 502
Osheroff, Douglas D., *2:* 232
Osmiridium, *2:* 272, 392; *3:* 502
Osmium (Os), *2:* **391–94**
Ovens, *3:* 488 (ill.)
Owens, Robert B., *3:* 479
Oxides, *2:* 401, 405
Oxygen (O), *2:* 395–407. *See also* under "Chemical properties" category in most entries
Oxygen tanks, *2:* 403 (ill.)
Oxyhemoglobin, *1:* 155
Oxyhydrogen, *2:* 252
Ozone, *1:* 77, 189–90; *2:* 400

P

P. *See* Phosphorus (P)
Pa. *See* Protactinium (Pa)
Pacemakers, *3:* 429 (ill.), 430, 435 (ill.), 436
Paint, *3:* 622
paint tubes, *1:* 12 (ill.)
Palladium (Pd), *3:* **409–13**
Panama, elements in, *2:* 218
Panchromium, *3:* 648
Pancreas, *2:* 266; *3:* 520
Paper materials, *3:* 622
Paracelsus, *1:* 34; *3:* 672–73
Paralysis, *2:* 266
Parathyroid gland, *3:* 520
Particle accelerators, *1:* 38, 49, 50 (ill.), 69 (ill.), 97, 157; *3:* 433, 439, 575–76, 626, 627 (ill.)
Patronite, *3:* 649
Pb. *See* Lead (Pb)
Pd. *See* Palladium (Pd)

Peligot, Eugène-Melchior, *3:* 640
Pencils, *1:* 103, 108
Pennsylvania, elements in, *1:* 55
Penny, *1:* 154, 154 (ill.)
Pens, *2:* 393 (ill.)
Pentlandite, *2:* 369
Perey, Marguerite, *1:* 191–93, 193 (ill.)
Periodic law, xxiv; *2:* 201, 210; *3:* 459, 511, 512
Periodic table, xxiv; *1:* 37–38; *2:* 201, 210; *3:* 459–60, 486, 511, 512
Peritoneum, *2:* 222–23
Perrier, Carlo, *3:* 576
Personality changes, *2:* 341
Peru, elements in, *1:* 61, 151; *2:* 301; *3:* 536, 612, 674
Pesticides, *1:* 77, 131
Petalite, *2:* 308, 309
Petroleum. *See* Oil entries
Philippines, elements in, *1:* 33
Phlogiston, *2:* 398
Phosphate rock, *3:* 419, 420–21
Phosphates, *2:* 401
Phosphors, *1:* 115–16, 177; *2:* 198, 290, 296, 338; *3:* 588, 668, 669
Phosphorus (P), *3:* **415–23**
Photocells. *See* Photovoltaic (solar) cells
Photoconductors, *3:* 519
Photocopying, *3:* 521, 522, 522 (ill.), 584
Photographic film, *2:* 265 (ill.)
Photography, *3:* 533, 537–38
Photosynthesis, *1:* 106; *2:* 274
Photovoltaic (solar) cells, *2:* 206–7, 259; *3:* 499, 522, 529, 594
Pica, *2:* 305
Piercing, *2:* 374, 375, 380
Pitchblende, *3:* 440, 468, 471–73, 478, 640, 641, 641 (ill.)
Plants, *1:* 72, 132
Plaster of paris, *1:* 88–89
Plastic materials, *3:* 622
Platinum (Pt), *2:* 270; *2:* 391–92, 393–94; *3:* **425–30,** 427 (ill.), 492, 502
Platinum family elements. *See* Iridium (Ir); Osmium (Os); Palladium (Pd); Platinum (Pt); Rhodium (Rh); Ruthenium (Ru)
Pliny, *1:* 20; *2:* 218, 279
Plumbum,, *2:* 300
Plunkett, Roy, *1:* 188
Plutonium (Pu), *3:* **431–37,** 434 (ill.)
Pm. *See* Promethium (Pm)
Po. *See* Polonium (Po)
Poda, Abbé Nicolaus, *3:* 496
Poison, *1:* 35, 36; *2:* 334, 340
Poland, elements in, *3:* 536
Pollucite, *1:* 121; *3:* 496
Polonium (Po), *3:* **439–42**
Polyhalite, *3:* 446
Polymer, *3:* 546
Polytetrafluorethylene (PTFE), *1:* 188
Polyvinyl chloride (PVC), *1:* 131
Portable X-ray machines, *1:* 16, 17; *3:* 661
Portugal, elements in, *3:* 635
Postage stamps, *1:* 177–78
Potash, *3:* 443–445, 447, 449
Potassium (K), *1:* 6; *3:* **443–51**
Potassium bromide, *1:* 75–76
Potassium chloride, *1:* 128
Potassium-dating, *3:* 446–47
Pr. *See* Praseodymium (Pr)
Praseodymium (Pr), *2:* 350; *3:* **453–57**
Precious metals. *See* Gold (Au); Platinum (Pt); Rhodium (Rh); Ruthenium (Ru); Silver (Ag)
Priestley, Joseph, *2:* 383, 395, 397 (ill.), 397–99; *3:* 654
Printers, *3:* 583 (ill.), 584
Promethium (Pm), *3:* **459–63**
Propylene oxide, *1:* 131–32
Protactinium (Pa), *3:* **465–70**
Protium, *2:* 248
Proustite, *3:* 536
Psilomelane, *2:* 329
Pt. *See* Platinum (Pt)
PTFE. *See* Polytetrafluorethylene (PTFE)
Pu. *See* Plutonium (Pu)
Puerto Rico, elements in, *3:* 563
Pulmonary system, *2:* 257
PVC. *See* Polyvinyl chloride (PVC)
Pyrargyrite, *3:* 536
Pyrites, *3:* 559, 563
Pyrochlore, *2:* 376–77
Pyrolusite, *1:* 126; *2:* 328, 329
Pyrrhotite, *2:* 369

Q

Quadramet, *3:* 508
Quartz, *2:* 222 (ill.)
Quicksilver, *2:* 333

R

Ra. *See* Radium (Ra)
Radiation, *1:* 106–7; *3:* 473

Radiation treatment, *3:* 475 (ill.)
Radiography, *1:* 46
Radium (Ra), *1:* 2; *3:* **471–76**
Radon (Rn), *3:* **477–83**
Radon gas, *3:* 475
Radon test kits, *3:* 480
Railroad tracks, *2:* 330 (ill.)
Ramsay, Sir William, *1:* 25, 26; *2:* 234, 287, 288–89, 355, 356–57; *3:* 478, 654, 655 (ill.), 657
Rare earth metals. *See* Cerium (Ce); Dysprosium (Dy); Erbium (Er); Europium (Eu); Gadolinium (Gd); Holmium (Ho); Lutetium (Lu); Neodymium (Nd); Praseodymium (Pr); Promethium (Pm); Samarium (Sm); Terbium (Tb); Thulium (Tm); Ytterbium (Yb)
Rare earth magnets, *1:* 70
Rayleigh, Lord, *1:* 25, 26
Rb. *See* Rubidium (Rb)
Re. *See* Rhenium (Re)
Realgar, *1:* 33
Réaumur, René Antoine Ferchault, *1:* 102
Reciprocating saw, *2:* 323 (ill.)
Rectifier, *3:* 529
Recycling, *2:* 302–3
Red phosphorus, *3:* 417, 418 (ill.), 423
Refractory materials, *1:* 71, 108, 139; *2:* 229, 318, 380; *3:* 531, 686
Reich, Ferdinand, *2:* 255, 256
Respiratory system, *1:* 57, 190; *2:* 394
Rf. *See* Rutherfordium (Rf)
Rh. *See* Rhodium (Rh)
Rhazes, *1:* 65
Rhenium (Re), *3:* **485–89,** 576
Rhodite, *3:* 492
Rhodium (Rh), *3:* **491–94**
Rhodizite, *1:* 121
Rhodochrosite, *2:* 329
Richardson, Robert C., *2:* 232
Richter, Hieronymus Theodor, *2:* 255, 256
Rn. *See* Radon (Rn)
Roasting, *2:* 302, 370–71; *3:* 487
Rock salt. *See* Salt
Rocket fuels, *2:* 403,
Rockets, *2:* 253
Rodenticides, *1:* 131; *3:* 591, 593–594
Rolla, Luigi, *3:* 460
Roscoe, Sir Henry Enfield, *3:* 648
Roscoelite, *3:* 649
Rose, Heinrich, *2:* 376; *3:* 570
Ru. *See* Ruthenium (Ru)
Rubber, *2:* 312, 374; *3:* 546, 564
Rubidium (Rb), *3:* **495–99,** 497 (ill.)
Russia, elements in, *1:* 9, 21, 55, 143, 151; *2:* 204, 212, 221, 257, 272, 280, 309, 321, 370, 389; *3:* 411, 419, 563, 635
Ruthenium (Ru), *3:* **501–4**
Rutherford, Daniel, *2:* 383
Rutherford, Ernest, *2:* 234, 234 (ill.), 235; *3:* 628
Rutherfordium (Rf), *3:* 626, 628, 630
Rutile, *3:* 619

S

S. *See* Sulfur (S)
Safety goggles, *3:* 456 (ill.)
Salt, *1:* 75–76, 128; *2:* 262, 266; *3:* 543, 551
Salt "domes," *1:* 128
Saltpeter, *2:* 401
Saltwater, *2:* 309
Samaria, *2:* 196
Samarium (Sm), *3:* **505–9**
Samarium-cobalt (SmCo) magnets, *3:* 508
Samarskite, *2:* 196; *3:* 506
Satellites, *2:* 213 (ill.), 275
Saturn (rocket), *2:* 404 (ill.)
Sb. *See* Antimony (Sb)
Sc. *See* Scandium (Sc)
Scaliger, Julius Caesar, *3:* 426
Scandium (Sc), *3:* **511–16**
Scheele, Carl Wilhelm, *1:* 41, 42, 125, 126, 127 (ill.), 185; *2:* 383, 395, 397, 399, *3:* 633, 634
Scheelite, *3:* 635
Schilling test, *1:* 144
Schmidt, Gerhard C., *3:* 597, 598
Schwanhard, Heinrich, *1:* 185
Se. *See* Selenium (Se)
Sea kelp, *2:* 264
Seaborg, Glenn, *1:* 15, 97, 157, 158 (ill.); *3:* 433, 628
Seaborgium (Sg), *3:* 626, 628, 630
Searchlights, *3:* 601 (ill.)
Seawater, *1:* 75–76, 128, 130; *2:* 263–64, 320–21; *3:* 543–44
Seaweed, *2:* 261–62, 264
Sefström, Nils Gabriel, *3:* 647, 648
Segrè, Emilio, *1:* 38; *3:* 576
Selenium (Se), *3:* 519 (ill.), 580, **517–23**

Tyrian purple, *1:* 74
Tyurin, V. A., *1:* 7

U

U. *See* Uranium (U)
Ukraine, elements in, *2:* 212, 329, 389
Ulexite, *1:* 67
Ulloa, Don Antonio de, *3:* 426
Ultraviolet (UV) radiation, *1:* 184, 190; *3:* 657
United Kingdom, elements in, *1:* 73; *2:* 212; *3:* 487
United States, elements in, *1:* 9, 43–44, 54, 61, 68, 73, 81, 83, 151, 163; *2:* 212, 221, 236, 280, 301, 309, 336, 345, 370, 389; *3:* 411, 419, 428, 514, 520, 536, 563, 666, 674
University of California at Berkeley (UCB), *1:* 49–50, 157; *2:* 351, 380; *3:* 431, 576, 626–27
University of Chicago, *1:* 15, 157
"Unripe gold," *3:* 580
Ununbiium (Uub), *3:* 626, 629, 630
Ununnilium (Uun), *3:* 626, 629, 630
Unununium (Uuu), *3:* 626, 629, 630
Uraninite, *3:* 641
Uranium (U), *1:* 38, 158–59; *2:* 234–35, 362; *3:* 434, 480, 598, 599, **639–46,** 642 (ill.)
Uranophane, *3:* 641
Uranus, *2:* 363; *3:* 640
Urbain, Georges, *1:* 170; *2:* 314; *3:* 659, 660
Urine, *3:* 415
Utah, elements in, *1:* 54, 151; *2:* 321, 336, 345; *3:* 419
Uub. *See* Ununbiium (Uub)
Uun. *See* Ununnilium (Uun)
Uuu. *See* Unununium (Uuu)

V

V. *See* Vanadium (V)
Vanadinite, *3:* 649
Vanadium (V), *1:* 172; *3:* **647–52,** 649 (ill.)
Vandermonde, C. A., *2:* 279
Vauquelin, Louis-Nicolas, *1:* 53, 54, 135, 136, 138
Vaz, Lopez, *2:* 218
Vegetable alkali, *3:* 444
Vermillion, *2:* 336
Vestium, *3:* 501
Virgin Islands (U.S.), elements in, *3:* 563
Virginium, *1:* 38, 192
Volcanoes, *2:* 351; *3:* 563
Vulcanization, *3:* 565, *3:* 583

W

W. *See* Tungsten (W)
Washington, elements in, *1:* 9; *2:* 321
Watches, *3:* 473
Water purification, *1:* 130
Weather balloons, *2:* 232 (ill.)
Welder's goggles, *3:* 456 (ill.), 457
Welding, *1:* 29; *2:* 237; *3:* 530
Welsbach, Baron von. *See* Auer, Carl (Baron von Welsbach)
Wetterhahn, Karen, *2:* 340
White phosphorus, *3:* 417, 423
Wicker, Henry, *2:* 318
Willemite, *3:* 673
Williamette (meteorite), *2: 278 (ill)*
Wilson's disease, *1:* 152–53, 156
Winkler, Clemens Alexander, *2:* 210, 212
Winthrop, John, *2:* 376
Wiring, *1:* 153
Witherite, *1:* 43
Wöhler, Friedrich, *1:* 6–7; *3:* 526
Wolfram, *3:* 633, 634. *See also* Tungsten
Wolframite, *3:* 514, 635
Wollaston, William Hyde, *3:* 409, 410, 491, 492, 493 (ill.)
Wood preservation, *1:* 36
Wood's metal, *1:* 82, 84
Wrought iron, *2:* 283–84
Wulfenite, *2:* 345
Wyoming, elements in, *2:* 236

X

X-ray diffraction analysis, *2:* 226
X rays, *1:* 43 (ill.), 44 (ill.), 46, 54, 68; *3:* 588 (ill.), 607–8
Xe. *See* Xenon (Xe)
Xenon (Xe), *3:* **653–57**
Xenotime, *2:* 196; *3:* 587

Y

Y. *See* Yttrium (Y)
YAG laser. *See* Yttrium-aluminum-garnet (YAG) laser
Yb. *See* Ytterbium (Yb)
Ytterbite, *3:* 506

Ytterbium (Yb), *2:* 314; *3:* **659–62**
Ytterby, Sweden, *1:* 161, 169–70; *2:* 195, 241, 313, 349–50; *3:* 586, 605, 606, 660, 664
Ytterite, *2:* 242
Yttria, *1:* 162, 170; *2:* 314, 350; *3:* 453, 454, 586, 606, 660
Yttrium (Y), *3:* 663–69, 665 (ill.), 667 (ill.)
Yttrium-aluminum-garnet (YAG) laser, *3:* 668
Yttrium garnets, *2:* 198 (ill.), 199

Z

Zaire, elements in, *1:* 143
Zambia, elements in, *1:* 143
ZBLAN, *2:* 297
Zimbabwe, elements in, *2:* 309
Zinc (Zn), *1:* 81, 82, 84; *2:* 202, 212; *3:* **671–79**
Zinc carbonate, *1:* 82
Zinc oxide, *3:* 675 (ill.)
Zincite, *3:* 673
Zircaloy, *3:* 683
Zircon, *2:* 228; *3:* 681, 684, 685–86
Zirconium (Zr), *2:* 225, 226–28, 229; *3:* **681–86,** 683 (ill.)
Zn. *See* Zinc (Zn)
Zr. *See* Zirconium